AF614699

METHODS IN MOLECULAR BIOLOGY™

For further volumes:
http://www.springer.com/series/7651

Corneal Regenerative Medicine

Methods and Protocols

Edited by

Bernice Wright and Che J. Connon

School of Pharmacy, University of Reading, Reading, UK

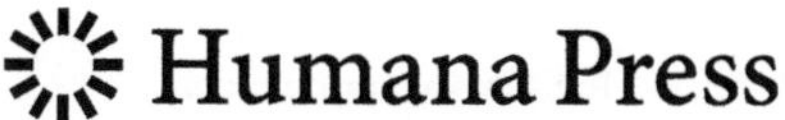

Editors
Bernice Wright
School of Pharmacy
University of Reading
Reading, UK

Che J. Connon
School of Pharmacy
University of Reading
Reading, UK

ISSN 1064-3745 ISSN 1940-6029 (electronic)
ISBN 978-1-62703-431-9 ISBN 978-1-62703-432-6 (eBook)
DOI 10.1007/978-1-62703-432-6
Springer New York Heidelberg Dordrecht London

Library of Congress Control Number: 2013937289

Printed on acid-free paper

Humana Press is a brand of Springer
Springer is part of Springer Science+Business Media (www.springer.com)

Preface

Continuous regeneration of the cornea is necessary to maintain this tissue in the transparent state that is essential for vision. Therapy for repair of the damaged anterior cornea is currently addressed through the transplantation of donor corneas or the delivery of limbal epithelial stem cells (LESC) to the ocular surface using amniotic membrane (AM) as a supporting scaffold. Certainly, stem cell therapy for ocular reconstruction is established and clinically routine. Research on the bioengineering of corneal equivalents as replacement tissue is also underway to develop viable corneal prosthetics.

This book is organized into sections describing the identification, characterization, and cultivation of LESC as well as the investigation of biopolymers used as the basis for corneal substitutes. The first part of the book sets the scene for the protocol sections, with a review of limbal epithelial cell (LEC) therapy and potential alternative cell and drug treatment modalities for an important cause of corneal blindness, limbal stem cell deficiency (LSCD) (Chapter 1). Advances in the construction of the artificial cornea are reviewed in Chapter 2. These reviews together with the protocol chapters clearly demonstrate the progression towards clinical translation of basic research approaches for understanding the cornea and addressing corneal dysfunction.

LESC are one of a very small group of progenitor cells that have fulfilled some of the enormous potential promised from stem cell therapy. In this book, we have described techniques for functional and biomarker assays that are critical for identification of LESC intended for clinical and research purposes. It has been strongly recommended (by scientists performing research on the cornea) that various distinct and complementary methodologies should be performed together for confident identification of LESC. In Chapter 3, a fundamental functional technique for LSC identification, clonal analysis, which enables classification of progenitor holoclones and differentiated meroclones and paraclones, is described. Chapter 4 details another functional assay for characterizing LESC, the bromodeoxyuridine incorporation assay. Protein markers are also used to indicate LESC phenotype. At present, the preferred strategy for the use of protein markers is to measure the levels of negative and positive protein markers for LESC and then use this to make a consensus decision about cell phenotype (Chapters 5 and 6). Western blotting is a powerful technique for this approach as it allows quantitative analysis of several proteins simultaneously (Chapter 6). Identifying protein markers together with functional assays (e.g., clonal assays) permits unequivocal characterization of LESC and is the best way to ascertain the phenotype of these cells.

We have described protocols for the culture of limbal epithelial (Chapter 7), stromal, and endothelial cells (Chapter 8), which emphasize the clinical and preclinical nature of these techniques. LEC are the cell type of choice for modern treatments applied to reversing corneal blindness caused by LSCD. We therefore, placed emphasis on the therapeutic management of these cells (i.e., methods for their preservation), and we have provided extensive notes discussing the manner that LEC cultivation impacts on their efficacy.

The potential of stromal keratocytes and endothelial cells for therapeutic application is just beginning to be recognized, so the protocol for the culture of those cells is basic and more appropriate for the laboratory.

The last part of this book describes protocols for tissue engineering approaches for corneal regeneration. Bioengineered prostheses are a viable option for repairing the injured cornea, due to a shortage of donor corneas and problems posed by a lifetime of immunosuppression necessary with the use of allogeneic replacement tissue. Hydrogels form major parts of the contemporary keratoprosthesis due to their excellent biocompatibility. Certainly, collagen gels are increasingly used as the key biomaterial component of KPro (Chapters 9 and 10). Importantly, we have discovered that collagen hydrogels preserve the stem phenotype of LESC, indicating this material as suitable for a bioartificial niche that may enhance corneal wound healing, by serving as both substitute tissue and an LESC delivery substrate in a prosthetic device. Collagen hydrogels were described as alternatives to AM, but other materials that can be used as replacements for AM include silk fibroin (Chapter 11) and nanofiber scaffolds (electrospun poly (lactide-co-glycolide) (Chapter 12) and copolymer Polyamide 6/12 (Chapter 13)). Silk fibroin is well tolerated when implanted into the cornea and is amenable to the addition of exogenous factors that promote cell attachment, growth, and differentiation. The structures of nanofibres can be controlled by electrospinning processes to produce scaffolds of distinct porosity and containing nanofibers with defined diameters, which determine their rate of degradation after transplantation. These substrates may, therefore, not only replace AM but also enhance therapeutic outcomes following the delivery of LESC.

Cell encapsulation technology is an integral part of structuring the corneal prosthetic, e.g., encapsulation of keratocyte feeder cells to ensure stratification of LEC cultured on a hydrogel surface. We believe that there is great potential for the future use of alginate gels (Chapter 14) within corneal substitutes, as these gels are established as scaffolds for cell encapsulation and tissue regeneration. Our own studies have demonstrated robust viability of LEC in alginate gels. The final chapter describes a protocol (Chapter 15) for in vivo imaging of the cornea following transplantation of biomaterials. Noninvasive methodologies for evaluating the efficacy of implanted corneal prosthetics are crucial to understanding the clinical usefulness of these medical devices. The method we describe allows clear clinical examinations of the integration of biomaterials into the damaged cornea and is at the cutting edge of diagnostic technologies applied to the cornea.

In this protocols volume, we provide a concise overview of essential techniques in the field of corneal regenerative medicine, highlighting novel strategies to guide the management of key therapies within this area of medicine. Collectively, the protocols that we have outlined for cultivation and characterization of corneal cells may be grouped into a robust toolbox of fundamental techniques that is useful for both the laboratory and the clinical setting.

Reading, UK

Bernice Wright
Che J. Connon

Acknowledgements

We kindly acknowledge all contributors involved in helping compile this book. For the extensive review of past, current, and potential future therapies for LSCD, as well as the LEC culture protocol chapter, we thank Dr. Tor Paaske Utheim and Professor Torstein Lyberg from Oslo University Hospital, Norway, and Dr. Sten Ræder from Stavanger University Hospital, Stavanger, Norway. Our thanks also go to Dr. Naresh Polisetti, Dr. Mohammad Mirazul Islam, Professor May Griffith from Linkoping University, Sweden, and Professor Nancy Joyce for the up-to-date perspectives review on the artificial cornea and the protocol chapter describing the culture of stromal and endothelial cells. We gratefully acknowledge the contribution of Dr. Neil Lagali, Professor Per Fagerholm and Professor May Griffith (Linkoping University, Sweden) for the protocol chapter on in vivo imaging of the transplanted artificial cornea. Our gratitude goes to Professor Ursula Schlötzer-Schrehardt for the protocol chapter describing the clonal analysis of LESC, a fundamental functional assay for the identification of these progenitor cells. For the bromodeoxyuridine incorporation assay protocol chapter, we thank Dr. Ashley Crane and Dr. Sanjoy Bhattacharya from the Bascom Palmer Eye Institute at the University of Miami, USA. We thank Dr. Martin Nakatsu and Dr. Sophie Deng (Doris Stein Eye Research Center, California, USA) who contributed their novel findings of the negative marker for LESC, stage-specific embryonic antigen-4, to producing a protocol chapter for enrichment of these cells from LEC populations. We thank Dr. Nicola Hunt (Newcastle University, UK) and Dr. Liam Grover (University of Birmingham, UK) for applying their expertise in 3D cell culture to the protocol chapter describing the encapsulation of cells in alginate hydrogels. We thank Dr. Pallavi Deshpande and Professor Sheila MacNeil (University of Sheffield, UK) for the Poly (lactide-co-glycolide) protocol that they are currently developing for the delivery of LEC to the cornea. For his use of nanofibre scaffolds to deliver both LESC and mesenchymal stem cells for the repair of the cornea, we thank Dr. Vladimir Holan (Institute of Molecular Genetics and Institute of Experimental Medicine, Charles University, Prague, Czech Republic). We are grateful to Dr. Damien Harkin and Dr. Laura Bray (Queensland Eye Institute, Brisbane, Australia) for compiling their preclinical studies on the use of silk fibroin as an alternative to AM into a very interesting protocol chapter. We included our own expertise to produce protocol chapters describing the identification of LESC by Western blotting and the construction of corneal equivalents using compressed collagen hydrogels, for which we thank Dr. Shengli Mi (Tsinghua University, China).

Contents

Contributors

SANJOY K. BHATTACHARYA • *Miller School of Medicine, University of Miami, Miami, FL, USA*
LAURA J. BRAY • *School of Biomedical Sciences, Queensland University of Technology, Brisbane, Australia*
TRAIAN V. CHIRILA • *Queensland Eye Institute, South Brisbane, Australia*
CHE J. CONNON • *School of Pharmacy, University of Reading, Reading, UK*
ASHLEY M. CRANE • *Miller School of Medicine, University of Miami, Miami, FL, USA*
SOPHIE X. DENG • *Jules Stein Eye Institute, UCLA, Los Angeles, CA, USA*
PALLAVI DESHPANDE • *The Kroto Research Institute, University of Sheffield, Sheffield, UK*
PER FAGERHOLM • *Division of Ophthalmology, Department of Clinical and Experimental Medicine, Linköping University, Linköping, Sweden*
KARINA A. GEORGE • *Queensland Eye Institute, South Brisbane, Australia*
MAY GRIFFITH • *Integrative Regenerative Medicine Centre, Department of Clinical and Experimental Medicine, Linköping University, Linköping, Sweden*
LIAM M. GROVER • *School of Chemical Engineering, University of Birmingham, Birmingham, UK*
DAMIEN G. HARKIN • *School of Biomedical Sciences, Queensland University of Technology, Brisbane, Australia*
VLADIMIR HOLAN • *Academy of Sciences of the Czech Republic, Institute of Experimental Medicine, Prague, Czech Republic*
NICOLA C. HUNT • *School of Chemical Engineering, University of Birmingham, Birmingham, UK*
MOHAMMAD MIRAZUL ISLAM • *Swedish Medical Nanoscience Center, Karolinska Institute, Stockholm, Sweden*
ELISKA JAVORKOVA • *Academy of Sciences of the Czech Republic, Institute of Experimental Medicine, Prague, Czech Republic*
NANCY C. JOYCE • *Emeritus Senior Scientist, Schepens Eye Research Institute, Boston, MA, USA; Department of Ophthalmology, Harvard Medical School, Boston, MA, USA*
NEIL LAGALI • *Division of Ophthalmology, Department of Clinical and Experimental Medicine, Linköping University, Linköping, Sweden*
TORSTEIN LYBERG • *Department of Medical Biochemistry, Oslo University Hospital, Oslo, Norway*
SHEILA MACNEIL • *The Kroto Research Institute, University of Sheffield, Sheffield, UK*
KIMBERLEY MERRETT • *Integrative Regenerative Medicine, Department of Clinical and Experimental Medicine, Linköping University, Linköping, Sweden*
SHENGLI MI • *Graduate School at Shenzhen, Tsinghua University, Shenzhen, People's Republic of China*
MARTIN N. NAKATSU • *Jules Stein Eye Institute, UCLA, Los Angeles, CA, USA*
NARESH POLISETTI • *Department of Ophthalmology, University of Erlangen-Nürnberg, Erlangen, Germany*

STEN RÆDER • *Center for Eye Research, Department of Ophthalmology, Ulleval University Hospital, University of Oslo, Oslo, Norway*

CHARANYA RAMACHANDRAN • *Sudhakar and Sreekanth Ravi stem cell laboratory, L V Prasad Eye Institute, Hyderabad, India*

VIRENDER S. SANGWAN • *Sudhakar and Sreekanth Ravi stem cell laboratory, L V Prasad Eye Institute, Hyderabad, India*

URSULA SCHLÖTZER-SCHREHARDT • *Department of Ophthalmology, University of Erlangen-Nürnberg, Erlangen, Germany*

SHUKO SUZUKI • *Queensland Eye Institute, South Brisbane, Australia*

PETER TROSAN • *Academy of Sciences of the Czech Republic, Institute of Experimental Medicine, Prague, Czech Republic*

TOR PAASKE UTHEIM • *Department of Medical Biochemistry, Oslo University Hospital, Oslo, Norway; Schepens Eye Research Institute, Massachusetts Eye and Ear, Department of Ophthalmology, Harvard Medical School, Boston, MA USA*

BERNICE WRIGHT • *School of Pharmacy, University of Reading, Reading, UK*

Part I

Perspectives

Chapter 1

Limbal Epithelial Cell Therapy: Past, Present, and Future

Tor Paaske Utheim

Abstract

The cornea, the clear window at the front of the eye, transmits light to the retina to enable vision. The corneal surface is renewed by stem cells located at the peripheral limbal region. These cells can be destroyed by a number of factors, including chemical burns, infections, and autoimmune diseases, which result in limbal stem cell deficiency (LSCD), a condition that can lead to blindness. Established therapy for LSCD based on ex vivo expanded limbal epithelial cells is currently at a stage of refinement. Therapy for LSCD is also rapidly evolving to include alternative cell types and clinical approaches as treatment modalities. In the present perspectives chapter, strategies to treat LSCD are discussed and advances in this important field of regenerative medicine are highlighted.

Key words Limbal stem cell deficiency, Limbal stem cell therapy, Limbal stem cells, Ex vivo expansion of limbal stem cells, Ocular surface reconstruction

1 The Cornea

The cornea is avascular and is supplied with glucose and oxygen, essential for maintenance of normal metabolic functions [1, 2], from the limbal circulation [3], the aqueous humor [4], and the tear film [1, 5]. It is heavily innervated with a density of nerve endings approximately 300–400 times greater than that in the skin [6]. From anterior to posterior, the cornea is composed of the epithelium, the Bowman's membrane, the stroma, the Descemet's membrane, and the endothelium (Fig. 1) [3].

The corneal epithelium is a non-keratinized, stratified squamous layer composed of a basal layer of column-shaped cells, a suprabasal layer of cuboid wing cells, and a superficial layer of flat squamous cells [7]. Differentiated squamous cells present microvilli (fingerlike projections) [8], which increase the cell surface area, allowing close association with the tear film and preventing desiccation of the corneal surface [3]. Tight junctions between epithelial cells enable the formation of a barrier on the corneal surface [9].

Bernice Wright and Che J. Connon (eds.), *Corneal Regenerative Medicine: Methods and Protocols*, Methods in Molecular Biology, vol. 1014, DOI 10.1007/978-1-62703-432-6_1, © Springer Science+Business Media New York 2013

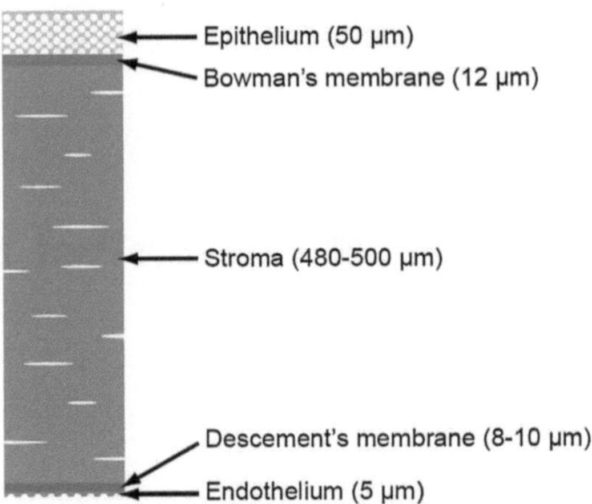

Fig. 1 Schematic of a transverse view of the cornea, illustrating the epithelium, Bowman's membrane, the stroma, Descemet's membrane, and the endothelium. Courtesy of Håkon Raanes, The Oslo School of Architecture and Design, Norway

Individual epithelial cells are connected to each other and to the basement membrane by desmosomes and hemidesmosomes, respectively. These structures are important in mediating cell migration in response to epithelial injury [10]. In vivo, the corneal epithelium is renewed every 9–12 months. Corneal transparency, which is essential for clear vision, is determined by (a) a smooth epithelium with no encroachment of conjunctival cells [11], (b) the absence of vasculature [3], (c) the unique architecture of the stroma, formed by uniformly spaced collagen fibers [12], (d) a functional endothelium that regulates corneal hydration [13], and (e) the production of crystalline proteins by keratocytes in the stroma [14].

The cornea is surrounded by the conjunctiva (Fig. 2), which rests on fibrovascular connective tissue containing blood vessels and lymphatics [15]. Langerhans cells, lymphocytes, melanocytes, and goblet cells are interspersed within the conjunctival epithelium [15]. Goblet cells produce mucin, which is essential for maintenance of the tear film and the integrity of the ocular surface [16]. Mucin deficiency has been implicated in several disabling diseases of the eye [17]. Abnormalities of the eyelids, especially the eye margins, can also lead to corneal pathology [18], which is most significant in the event of preexisting corneal disease [19]. Tears are produced and drained continuously to maintain the corneal and conjunctival epithelium in a moist state. The tear film provides bacteriostatic components, oxygen, and nutrients to the cornea [20].

2 Limbal Stem Cells

Corneal integrity and function are dependent on the self-renewing properties of the corneal epithelium [3]. This process relies on a small population of putative limbal stem cells (LSC) that are located

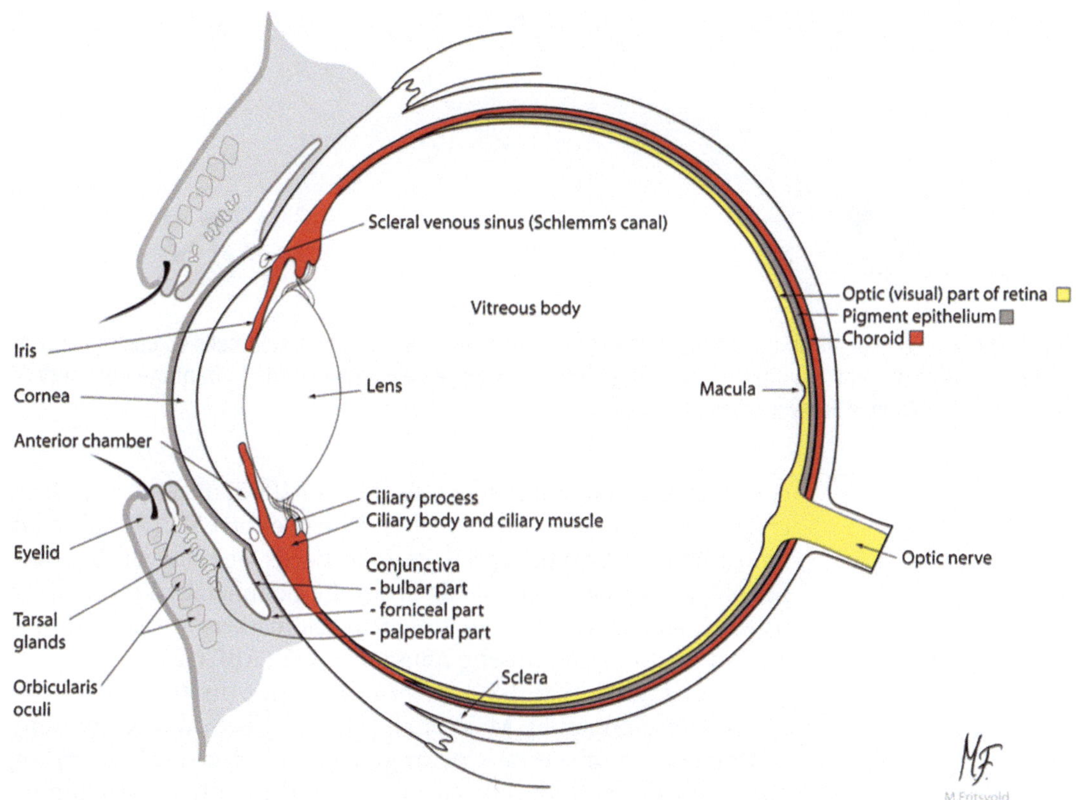

Fig. 2 The anatomy of the human eye. Courtesy of Dr. Magnus Fritsvold, Oslo, Norway

in the basal region of the limbus [21, 22]. LSC self-renew and give rise to fast-dividing, transit amplifying cells (TAC) [23]. TAC undergo a limited number of divisions before they become terminally (post-mitotic) differentiated cells (TDC) [24]. The process of TDC shedding from the ocular surface during normal wear and tear stimulates epithelial cell division, migration, and differentiation [25]. Epithelial cell loss (Z) from the corneal surface has been defined in the "X, Y, Z hypothesis" as the sum of the proliferation of basal cells (X) and the centripetal migration of cells (Y) (Fig. 3) [26]. Growth factors play important roles in the maintenance and wound healing of the cornea, and are supplied by the adjacent tear film [27], the aqueous humor [28], the epithelial cells [29], and keratocytes in the supporting stroma [30].

The LSC niche is located at the palisades of Vogt [21], a microenvironment that consists of cellular and extracellular components, hypothesized to regulate the fate of these cells [31, 32]. The palisades of Vogt are in close proximity to blood vessels [33], which provide nutrients and oxygen to LSC, and these undulating structures protect LSC from shearing forces [34]. The pronounced heterogeneity of basal membrane proteins at the corneal-limbal transition zone may provide a unique microenvironment for corneal

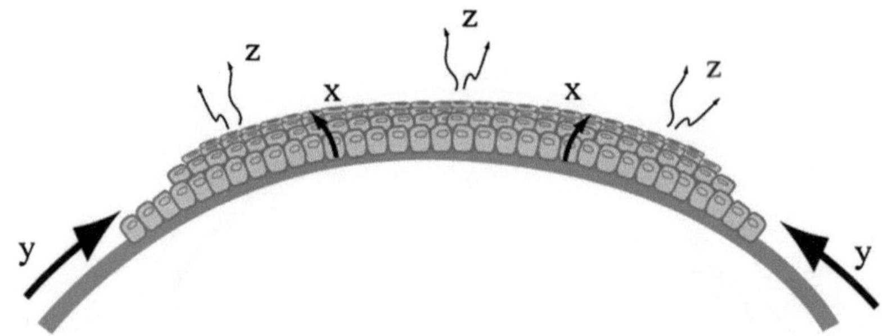

Fig. 3 The X, Y, Z hypothesis of corneal epithelial maintenance. The desquamated cells (Z component) are continuously replaced by the basal cells (X) that divide as well as cells that migrate in from the periphery (Y). Courtesy of Dr. Magnus Fritsvold, Oslo, Norway

epithelial progenitor and stem cells [35]. In 2005, the limbal epithelial crypt was reported as a potential novel niche for LSC [36, 37]. The limbal epithelial crypt represents solid cords of cells which extend peripherally or circumferentially from the peripheral end of a limbal palisade, and express the putative stem cell markers cytokeratin (CK) 14 and ATP-binding cassette transporter G2 (ABCG2).

The dogma that LSC exclusively reside in the limbus was recently challenged by Majo et al. [38] as cells from the uninjured central cornea of several species generated holoclones, presenting characteristics of stem cells. In support of the controversial findings reported by Majo et al. [38], Dua and colleagues [39] discovered normal central islands of corneal epithelium in eight eyes with clinically apparent total limbal stem cell deficiency (LSCD). These observations may have two explanations: (a) although not clinically visible, some LSC remain and thereby contribute to the maintenance of the central epithelium, or (b) basal cells of the central surviving epithelium are independently capable of maintaining the central epithelium.

The location of stem cells throughout the ocular surface reported by Majo et al. (2008) may be explained by (a) differences in the cornea of adult and infant humans (the two human corneas investigated by Majo et al. (2008) were from 1- and 4-year-old children, whereas other studies have involved adults) [38], and (b) differences in the anatomy of the cornea between species [40]. The avascular Bowman's layer underlying the central corneal epithelium is present in humans and primates, but is not present in other mammals [40]. The study by Majo et al. (2008) is inconsistent with many known growth, differentiation, and cell migration properties of the anterior ocular epithelia [41].

2.1 Clinical Indications for Limbal Stem Cells

The limbus is a common region for certain congenital anomalies to occur, such as limbal dermoids with ectopic brain [42] or bony tissue [43], which suggest the presence of undifferentiated cells. The first clinical indication for the presence of stem cells in the limbus was the observation of melanin moving from this region towards an epithelial defect, following wounding of rabbit corneas [44]. Hanna [45] documented a centripetal movement of corneal epithelial cells from the limbus to the central cornea, indicating that proliferating precursor cells were located at the limbus. These findings were confirmed in later studies [21, 46, 47].

Surgical indications for LSC were demonstrated when donor epithelium gradually replaced host epithelium following lamellar keratoplasty [48]. Srinivasan and Eakins [49] showed that it is impossible to create permanent corneal epithelial defects in laboratory animals without damaging the limbus. The LSC concept has been supported by clinical observations of abnormal corneal epithelial wound healing after removal of the limbal epithelium [50–52]. Furthermore, the high success rate of LSC transplantation strengthens the possibility that stem cells are located in the limbal region of the cornea.

2.2 Limbal Stem Cell Identification

The controversy of the location of LSC can be abated through specific identification and characterization of these cells. This is not, however, a trivial task as, to date, there is no definitive marker for LSC. LSC are currently identified by several indirect methods, including investigation of morphology, presence/absence of phenotypic markers (*see* Chapter 6), clonogenicity (*see* Chapter 3), and label retention (*see* Chapter 4).

Basal cells of the limbal epithelium express features of immature cells, including small cell size with a cytoplasm rich in tonofilaments, a high nuclear-cytoplasmic ratio, euchromatin-rich nuclei, and scarcely detectable nucleoli [53–55]. Numerous markers for LSC have been proposed, but there is currently no consensus on a specific marker for these cells. The phenotypic description of LSC currently relies on the combination of positive expression of putative stem cell-associated markers and negative expression of differentiation-associated markers [56]. Key putative phenotypic markers for LSC include ABCG2 and ΔNp63α; and common markers for differentiated limbal epithelial cells (LEC) include CK-3 and CK-12. ABCG2 is expressed by discrete clusters of basal LEC, with no expression being detected in the corneal epithelium [56–60]. The ΔNp63α isoform of the transcription factor p63 is present in approximately 8 % of the total cell population in the basal layer of the limbus, and is a more specific marker for LSC than p63 [61]. CK-3 and CK-12 are present in the suprabasal layers of the limbal epithelium [24, 62] and are absent from the basal limbal layer [24, 63–65].

Clonal analysis, a method for investigating the growth potential of cells, has indicated that stem cells are located in the limbus [66, 67]. LSC can also be identified by the retention of DNA labels as they are slow-cycling and only divide occasionally [68]. Chasing an initial pulse dose of DNA labels, Cotsarelis et al. [22] found slow-cycling label-retaining cells in the limbal basal epithelial region of the mouse cornea. The more differentiated and more rapidly dividing daughter TAC undergo dilution of the label through multiple divisions. The slow-cycling nature of LEC was subsequently confirmed in animal models [69–71].

3 Limbal Stem Cell Deficiency

Any process or disease (Table 1) that results in a reduced number of LSC and/or impaired function of these cells may produce clinically apparent LSCD. In the majority of cases, the etiology is thought to be a combination of reduced number and function of LSC [72]. Depending on the extent of the disorder, LSCD is classified as either partial or total. If the pupillary area of the eye is covered by encroaching conjunctival tissue, intervention is usually required [73].

The incidence and prevalence of LSCD worldwide are unknown, but the prevalence in India is estimated to be approximately 1.5 million [74] and the incidence in North America is approximated to be "thousands" [75]. Schwab and Isseroff [75] suggested that the global incidence of blindness from LSCD is much higher because of the many cases of trachoma (Fig. 4) and the greater incidence of ocular trauma outside North America. Chemical burns are, however, the most frequent indication for transplantations using ex vivo cultured LEC.

3.1 Clinical Presentation of Limbal Stem Cell Deficiency

Symptoms of LSCD include irritation, epiphora, blepharospasms, photophobia, recurrent episodes of pain (epithelial breakdown), and decreased vision [76]. Clinical findings of LSCD include a loss of the palisades of Vogt, opaque corneal epithelium with variable thickness and a dull and irregular reflex, punctate epithelial keratopathy, epithelial defects, abnormal staining with fluorescein, unstable tear film, superficial and deep corneal neovascularization, and fibrovascular pannus (Fig. 5) [46, 72, 73, 74, 76–78].

Healthy LSC act as a barrier against invasion of the cornea by conjunctival tissue [79, 80]. In LSCD, however, conjunctival tissue migrates towards the central cornea in a process called conjunctivalization, which is considered the hallmark of LSCD [81]. Chronic inflammation not only leads to the death of more LSC but also leaves surviving stem cells unable to function properly, explaining the worsening of clinical symptoms over time [72]. Persistence of epithelial defects may lead to ulceration, and eventually even corneal perforation [72]. In patients with severe dry eyes suffering from LSCD, the epithelium becomes partially or totally keratinized [73]. The end stage of LSCD, irrespective of cause, is scarring and ultimately calcification. At this stage, inflammation has normally subsided, and the eye is relatively comfortable.

3.2 Diagnosis of Limbal Stem Cell Deficiency

The diagnosis of LSCD is often possible clinically. Sophisticated testing is not necessary for patients with conditions known to cause LSCD (e.g., chemical injuries and Stevens–Johnson syndrome)

Table 1
Etiological classification of limbal stem cell deficiency

I. Idiopathic [76–81]
II. Hereditary
Anirida [78, 80–83], including the congenital aniridia variant: minimally abnormal irides with severe LSCD [84]
Dominantly inherited keratitis [85]
Autosomal dominant keratitis [86]
Multiple endocrine neoplasia [11, 87]
Ectrodactyly-ectodermal dysplasia-clefting syndrome [88, 89]
Ectodermal dysplasia [78, 83, 90]
Keratitis–ichthyosis–deafness (KID) syndrome [91, 92]
Xeroderma pigmentosa [93]
Lacrimo-auriculo-dento-digital (LADD) syndrome (Levy-Hollister syndrome) [94, 95]
Autoimmune polyendocrinopathy-candidiasis-ectodermal dystrophy (APECED) [96, 97]
Bilateral LSCD with chromosomal translocation of 3p and 9p [98]
Epidermolysis bullosa [80] and its subgroup epidermolysis bullosa dystrophica hallopeau-Siemens [99]
Iris coloboma [100]
Dyskeratosis congenita [101]
Gelatinous drop-like dystrophy [102]
III. Trauma
Chemical or thermal burns [77, 78, 80, 81, 83, 87, 103–132]
Delayed-onset mustard gas keratitis [133–135]
Perforating eye injury [80]
IV. Eye disease
Pterygium and pseudopterygium [80, 117, 126, 136–139]
Infections (including trachoma) [75, 87, 113, 140–142]
Keratitis [77], including neurotrophic keratitis [11] and peripheral ulcerative keratitis (Mooren`s ulcer) [143]
Keratoconjunctivitis sicca [144], atopic keratoconjunctivitis [143], vernal keratoconjunctivitis [145, 146], and cicatricial keratoconjunctivitis [127]
Chlamydia conjunctivitis [80]
Bullous keratopathy [147]
Corneal intraepithelial dysplasia [148]
Sclerocornea [149]

(continued)

Table 1
(continued)

Ocular tumors [126, 144], including recurrent ocular surface melanoma [150]
Limbal dermoid [151]
Salzmann nodular corneal dystrophy [81]
V. Iatrogenic (condition inadvertently caused by medical treatment)
Multiple ocular surface operations [77, 152–154]
Tumor excision [80]
Multiple intravitreal injections [155]
Cryotherapy on the ocular surface [11]
Phototherapeutic keratectomy [156, 157]
Radiotherapy [113, 158, 159]
Local chemotherapy (antimetabolites, e.g., 5 fluorouracil (5FU) [160] and mitomycin C (MMC) [80, 161–163]), and systemic chemotherapy [164, 165]
VI. Systemic Diseases
Diabetes [166]
Vitamin A deficiency [144]
Graft-versus-host disease [80, 81, 167]
Rosacea [83]
VII. Autoimmune Diseases
Stevens–Johnson syndrome [11, 81, 83, 108–110, 116, 121, 127, 168, 169] Toxic epidermal necrolysis [170]
Ocular cicatricial pemphigoid [81, 83, 109, 127, 171, 172]
Autoimmune polyendocrinopathy [173]
VIII. Others
Contact lens use [174]
Ultraviolet radiation [175]

with typical clinical signs, including conjunctivalization, loss of the palisades of Vogt, and persistent epithelial defects [72]. A tissue diagnosis is, however, warranted for patients with less clear symptoms, especially if LSC transplantation is being considered as a therapy. Under those circumstances, impression cytology [77–79, 87, 104, 105, 108, 112, 120, 122, 127, 169, 182] or in vivo microscopy (*see* Chapter 15) [78] can be of help in making the diagnosis.

For impression cytology, a cellulose acetate filter [183, 184] or a polytetrafluoroethylene membrane [185] may be pressed onto the corneal surface under topical anesthesia to obtain superficial epithelial cells (Fig. 6). Commonly used markers for conjunctivalization include the presence of goblet cells [186] and CK-19 [112, 187], whereas

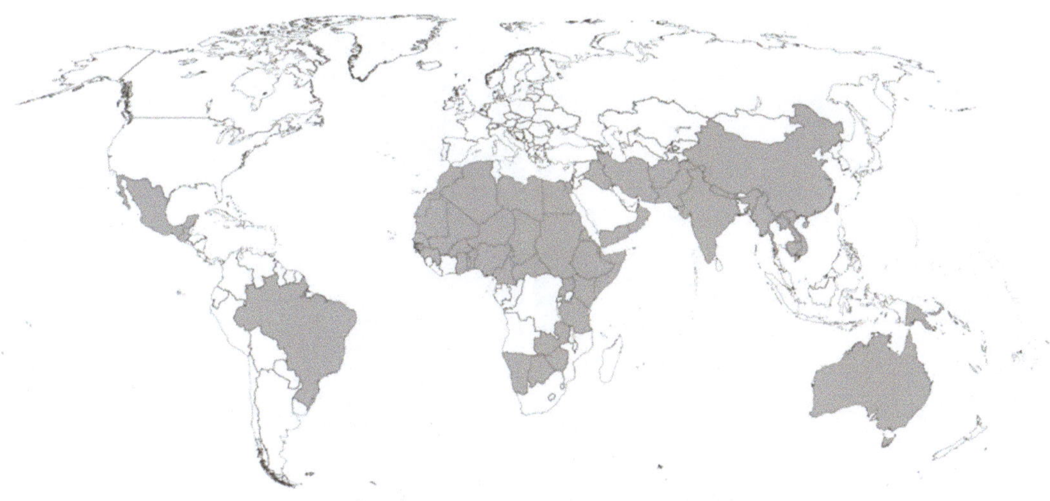

Fig. 4 Map of trachoma endemic countries in 2009. Courtesy of Håkon Raanes, The Oslo School of Architecture and Design, Norway

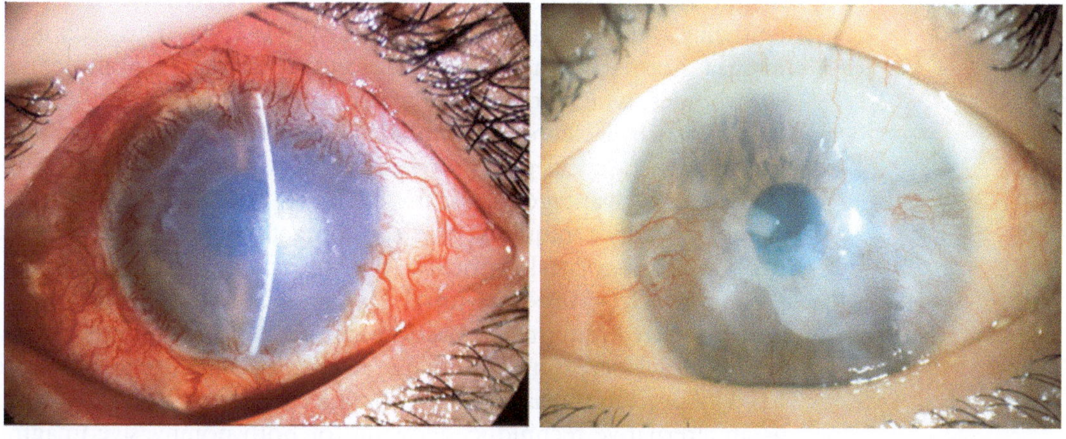

Fig. 5 Limbal stem cell deficiency. Courtesy of Dr. Takahiro Nakamura, Department of Ophthalmology, Kyoto Prefectural University of Medicine, Japan

mucin 1 [188], CK-7 [189], and CK-13 [190] are more specific markers for conjunctival epithelial cells. Sangwan et al. [145] demonstrated the absence of goblet cells in one-third of patients with clinical diagnosis of LSCD. These patients later received transplantations of ex vivo cultured LEC, resulting in restoration of the ocular surface. Therefore, whilst goblet cells confirm the diagnosis of LSCD, the absence of these cells does not exclude the possibility of this disorder. Interestingly, Rama et al. [113] stopped performing impression cytology during a large study as they found the additional information not necessary for making the diagnosis.

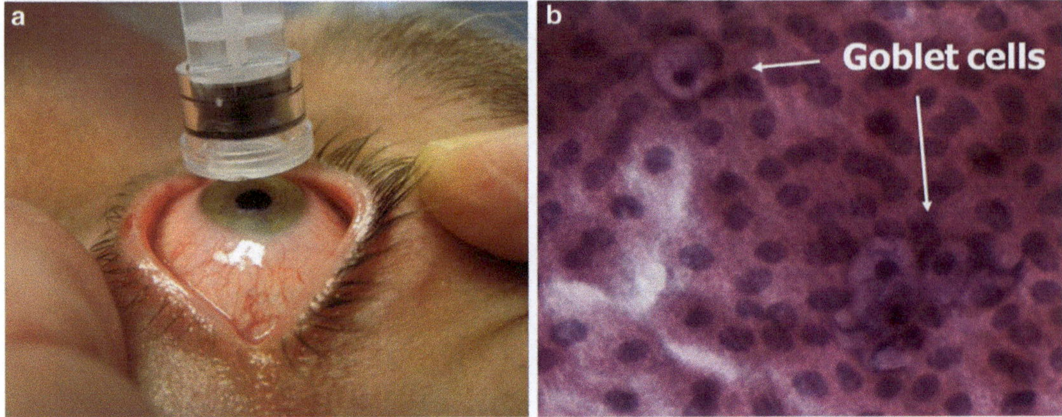

Fig. 6 Impression cytology. (**a**) Superficial epithelial cells are noninvasively harvested from the ocular surface using a polytetrafluoroethylene membrane mounted on the end of a plastic syringe plunger ad modum Dr. David Hugo Engelsvold, Department of Ophthalmology, Stavanger University Hospital, Norway. (**b**) Goblet cells scattered among epithelial cells

In vivo confocal microscopy has emerged as a promising diagnostic tool for LSCD as it does not involve removal of corneal epithelial cells for subsequent analyses [191–194]. This technique images normal corneal epithelial cells with bright well-defined membranes, which contrast to hyper-reflective conjunctival epithelial cells with ill-defined membranes [39]. Although the quality of images generated using in vivo confocal microscopy is impressive, there are disadvantages to the use of this approach: (a) the measurement procedure requires physical contact between the imaging probe and the corneal epithelium or conjunctiva, and (b) the field of view is small (~200 × 200 μm). This implies that a 2D view of a 1 × 1 mm area requires acquisition and overlap of multiple images, significantly increasing the image acquisition and processing time [195]. Spectral domain optical coherence tomography has emerged as an alternative technique, allowing for both noninvasive imaging of the ocular surface and a larger field of view at the expense of the resolution [195].

4 Treatment Approaches for Limbal Stem Cell Deficiency

4.1 Historical Overview

A wide array of cell-based strategies to treat LSCD have been proposed over the past 62 years (Table 2), including amniotic membrane (AM), representing the first cell-based procedure in 1940 [196], and transplantation of conjunctival-limbal-corneal epithelium, marking the first LSC transplantation in 1965. Standard therapy for corneal repair, penetrating keratoplasty, which replaces the center of the cornea with a donor corneal button, is not a viable option for patients suffering from LSCD as this procedure does not restore the stem cell population [144, 197, 198].

Table 2
Milestones in cell-based therapy for limbal stem cell deficiency

Year	Source of tissue transplanted onto the ocular surface	Species	References
1940	AM with chorion	Humans	[196]
1946	AM without chorion	Humans	[199]
1965	Conjunctival-limbal-corneal epithelium	Humans	[200]
1977	Conjunctival autograft (CAU)	Humans	[201]
1984	Corneal lenticules (keratoepithelioplasty)	Humans	[202]
1989	Conjunctival limbal autograft (CLAU)	Humans	[203]
1990	Keratolimbal allograft (KLAL)	Humans	[204]
1995	Living-related conjunctival limbal allograft (lr-CLAL)	Humans	[205]
1995	Living-related conjunctival allograft (lr-CAL)	Humans	[206]
1997	Ex vivo cultivated limbal autograft (EVLAU)	Humans	[207]
1999	Homologous penetrating central limbo-keratoplasty (HPCLK)	Humans	[208]
2000	Limbal epithelium cultured on AM	Humans	[117]
2003	Ex vivo cultivated oral mucosa autograft (EVOMAU)	Rabbits	[209]
2004	Ex vivo cultivated oral mucosa autograft (EVOMAU)	Humans	[210]
2004	Cultured embryonic stem cells (ESC)	Mice	[211]
2006	Cultured conjunctival epithelial cells	Rabbits	[212]
2006	Cultured bone marrow-derived mesenchymal stem cells (BMMSC)	Rats	[213]
2007	Cultured epidermal adult stem cells (EpiASC)	Goats	[214]
2009	Cultured immature dental pulp stem cells (IDPSC)	Rabbits	[215]
2009	Ex vivo cultivated conjunctival autograft (EVCAU)	Humans	[150]
2010	Cultured hair follicle bulge-derived stem cells (HFSC)	Mice	[216]
2011	Cultured umbilical cord-lining stem cells	Rabbits	[218]

Historical overview of tissue (cultured and non-cultured) that has been transplanted onto the ocular surface with the purpose of treating conditions associated with LSCD, including attempts before the concept of LSCD was established. Text highlighted in dark grey indicates transplantation of non-cultured tissue, and light grey-highlighted text shows transplantation of cultured tissue

4.2 Overview of Treatment Strategies for LSCD

The strategies for treating LSCD can be broadly divided into three categories: (a) transplantation of cultured cells, (b) transplantation of non-cultured cells, and (c) alternative approaches, which include implantation of keratoprostheses, repeated debridement of conjunctival tissue, bandage contact lenses, oxygen therapy, steroids, anti-angiogenic medication, lubricating eye drops, autologous serum, AM extract, and limbal fibroblast conditioned medium. There are potentially several benefits to using cultured tissue, for example the need for only a small biopsy, but also disadvantages, including high costs. By using non-limbal tissue, autologous transplantation is possible in cases of bilateral LSCD, circumventing issues related to allografts, such as transmission of microorganisms and graft rejection.

4.3 Preoperative Considerations of Cell-Based Therapy for Limbal Stem Cell Deficiency

Four main factors threaten the success of any cell-based procedure for treating LSCD: (a) severe tear film deficiency [219, 220]; (b) uncontrolled inflammation [221]; (c) severe lid abnormalities, such as lagophthalmos, misdirected lashes, and malpositioned or keratinized lid margins [222]; and (d) corneal and conjunctival anesthesia [220]. These conditions should be investigated and, if possible, addressed prior to surgery, although successful limbal transplantations were previously reported in spite of diminished tear film [81] and other suboptimal conditions [223].

For the transplantation of allografts, severe inflammation is a particular concern. Although the exact mechanism remains unclear, inflammatory cytokines, such as interferon-γ, can up-regulate Fas and human leukocyte antigen (HLA) class II, resulting in apoptosis of the epithelium [221, 224]. Up-regulation of HLA class II due to inflammation may enhance immune sensitization and lead to allograft rejection [225].

In a retrospective study of patients undergoing various forms of LSC transplantation, Desousa et al. [83] observed that concurrent adnexal abnormalities were associated with poor graft outcome after transplantation. In this study, corneal exposure with lagophthalmos was the most frequently encountered problem. Fornix involvement was associated with 50 % risk of failure. Furthermore, the authors suggested that stem cell grafting may exacerbate adnexal disease, and concluded that surgery to correct eyelid malpositioning and closure is important before and, if necessary, after LSC transplantation.

Decreased tear production causes desiccation and nutritional deprivation of ocular surface epithelia. Patients with dry eyes have a lower likelihood of success following limbal transplantations than those with normal tear production [219, 226]. Keratinization, a consequence of decreased tear production, may also reduce the likelihood of successful limbal transplantation [178, 226].

Another preoperative consideration is the choice of cultured or non-cultured tissue and cell type. In 2007, Ang and associates [170] compared the effects of transplantation of cultivated LEC and "conventional LSC transplantation" in a 32-year-old man with Stevens–Johnson syndrome and bilateral total LSCD. Transplantation of ex vivo expanded LEC resulted in better vision and clinical results with less stromal scarring during a 4-year follow-up. Moreover, complete corneal epithelialization was achieved 48 h after transplantation with cultivated LSC compared with 3 weeks after "conventional LSC transplantation" [170]. A larger study is warranted to draw conclusions from this highly interesting subject. Another important study was initiated by Ang and colleagues [227], demonstrating that cultured human conjunctival epithelial and cultured LEC produce clinically equivalent outcomes. Cell sources such as hair follicles [216], epidermis [214], conjunctiva [150], limbus [207], and oral mucosa [209], which are more easily accessible than dental pulp [228], bone marrow mesenchymal stem cells (MSC) [213], and embryonic stem cells (ESC) [229], may be the cell types of choice, if other cells that are more effective at treating LSCD are not found.

4.4 Transplantation of Cultured Cells to Treat Limbal Stem Cell Deficiency

4.4.1 Cultured Limbal Epithelial Cells

Ex vivo expansion of LEC was reported for the first time in 1997 by Pellegrini and coworkers [207]. The principle of this method is to culture LEC harvested from the patient [207], a living relative [126], or a cadaver [109] on a substrate [116, 150, 207] in the laboratory and transfer the cultured tissue onto the eyes of patients suffering from LSCD. The surgical procedure to treat LSCD with cultured LEC involves removal of the fibrovascular pannus before the substrate supporting these cells is attached onto the bare cornea. In cases where penetrating keratoplasty is required, the procedure should be performed at least 6 weeks after the transplantation of cultured LEC [230]. Repeating the transplantation of cultured LEC in the same eye has been demonstrated to increase the final clinical success rate from 68.5 to 81.8 % [113] and to result in improved visual acuity (discerning two lines or more of symbols on an eye chart) in 76 % (38/50 eyes) of the patients [231].

The method for transplantation of cultured LEC has several advantages. It enables reconstruction of the ocular surface from a small biopsy of healthy eye, hence minimizing the risk of stem cell failure in the donor eye [232]. This allows a further biopsy to be obtained, if necessary [78, 110, 113, 120, 121, 126, 182, 231]. Many close relatives will be amenable to a small biopsy, but not to a large one (e.g., 90°) with the known associated risks. The high rate of limbal allograft rejection is associated with the presence of antigen-presenting cells (APC) [233], vessels, and lymphatics in the limbal area [234]. In contrast to procedures that directly transplant limbal tissue, very few [234] or no APC [235] are contained in cultivated corneal epithelial sheets, hence reducing the risk of immune responses when allografts are used [236].

The main disadvantages associated with the cultivation of LEC are costs related to the establishment and maintenance of a stem cell laboratory. Miri et al. [237] estimated that a single cultivated limbal epithelial transplantation costs approximately €12,000. Moreover, the process is labor intensive (approximately 2 weeks of culture is needed) and requires expertise in cell culture.

In 2007, Shortt et al. [223] found that the overall success rate of ex vivo expansion of LEC was 75.7 % (among the 17 clinical studies identified, 147 out of 194 eyes were successfully treated) [81, 107, 109–112, 114–117, 125, 126, 171, 207, 238–240]. In 2011, Baylis and associates [241] in reviewing 28 case reports and series published over the past 13 years registered an almost similar success rate, i.e., 76 % (443/581 eyes) [77, 78, 80, 81, 104–117, 125–127, 150, 169–171, 182, 207, 240]. In the present review, 51 clinical reports, including sub-studies, addressing cultured LEC were identified (until February 2012) [77–81, 83, 87, 99, 103–128, 145, 150, 169–171, 182, 207, 230, 231, 238–240, 242–246]. These studies were performed in centers in Australia [119, 150], Brazil [79], China [118], Germany [80, 107, 169, 243, 244], Great Britain [78, 83, 182, 240], India [103, 106, 114, 115, 122, 125, 128, 145, 246], Iran [104], Italy [77, 105, 112, 113, 120, 207], Poland [123], the USA [87, 116, 126], Taiwan [117], and Japan [81, 108–111, 121, 124, 127, 170, 171]. Four of these studies focused on the characterization of transplanted sheets of cultured LEC [121, 242–244]. The success rate of transplantation of cultured LEC is, however, similar to previous reports, i.e., 74.4 % (661/889 eyes). The success of autografts was 75 % (579/772 eyes), whereas the success rate of allografts was 70.1 % (82/117 eyes). These percentages were calculated after excluding sub-studies and eyes due to the lack of data on success rate.

Approximately half of the eyes receiving cultured limbal epithelial transplants displayed visual improvement that allowed patients to discern two lines or more of symbols on an eye chart. In cases of partial LSCD, success and visual acuity may not correlate well. Similarly, if there is stromal scarring, despite restored epithelium and a marked difference in symptoms, there may be no significant visual improvement. In most studies, stromal scarring was not systematically recorded, but could be assessed indirectly by the number of patients who had simultaneous (25 eyes [80, 109, 110, 113, 116, 230]) or subsequent corneal transplants (322 eyes [77, 80, 104–108, 112–116, 119, 120, 124, 127, 182, 207, 230, 240]).

Among the current clinical studies involving cultured LEC, nine studies had a mean follow-up time of less than 1 year [78, 87, 109, 122, 123, 126, 150, 171, 246], 15 studies had a mean follow-up of 1–2 years [79, 81, 105, 107, 110–112, 114–117, 119, 125, 128, 182], and 14 studies had a mean follow-up of more than 2 years [77, 80, 104, 106, 108, 113, 120, 124, 127, 145, 169, 170, 231, 240]. The longest follow-up period presented was 10 years [113]. The most common etiologies of LSCD included in these studies were ocular burns (chemical and thermal) (~ 700 eyes) [77, 78, 80, 81, 83, 87, 103–128], Stevens–Johnson syndrome (40 eyes) [81, 83, 108–110, 116, 121, 127, 170, 171], pterygium or pseudopterygium (18 eyes) [80, 117, 126], and aniridia (12 eyes) [78, 80, 81, 83]. All other causes of LSCD constituted less than ten eyes. Complications observed after the grafting of cultured LEC in these studies included bleeding (77 eyes) [113, 120, 128], inflammation (67 eyes) [113, 120], blepharitis and epitheliopathy (35 eyes) [113], keratitis (12 eyes) [78, 81, 103, 113, 115, 116, 120, 128, 240], residual fibrin (11 eyes) [113], rejection of cultured LEC (9 eyes) [109, 114, 122], corneal perforation (7 eyes) [80, 104, 127, 169], corneal ulceration (5 eyes) [120, 127], sterile corneal melts (5 eyes) [128], glaucoma or increased intraocular pressure (4 eyes) [114, 115, 120], infections (4 eyes) [109, 127], phthisis (2 eyes) [115], descemetocele (1 eye) [120], pyogenic granuloma (1 eye) [116], epithelial loss (1 eye) [116], pseudoepitheliomatous hyperplasia [246], and graft detachment (1 eye) [78].

Follow-up time should ideally be at least two years as failures typically occur during this period, and especially the first year [113, 128, 230]. In the largest clinical study (200 eyes) involving cultured LEC, Sangwan and colleagues [128] noticed that all failures occurred within 20.6 months, and 81 % (47/58 eyes) of failures occurred within the first year. After repeated transplantations, the same group observed that 88 % (15/17 eyes) of failures occurred within the first year [231].

The Fate of Transplanted Limbal Cells Used to Treat Limbal Stem Cell Deficiency

Studies on LEC survival after transplantation for the treatment of LSCD are contradictory. Methods used to study the survival of allogeneic transplanted LEC include (a) fluorescent in situ hybridization of the sex chromosome [247], (b) DNA fingerprinting [248–250], and (c) xenotransplantation studies [251]. In vivo tracking of autologous transplanted cells are, however, problematic and whether or not labeling adversely affects the characteristics or function of these cells has yet to be proven.

Daya et al. [240] demonstrated that ex vivo expanded donor cell DNA was not present on the ocular surface 9 months after transplantation of cultured LEC. Williams et al. [249] detected donor cells only up to 12 weeks after transplantation of limbal tissue. In other studies involving limbal allo-transplantations, donor cells were detected after approximately 2.4 years [252], 2.5 years [253], and 3.5 years [254, 255]. Studies on the survival of donor cells are important to address the need for immunosuppression after LSC transplantation.

Mechanisms for Reconstruction of the Ocular Surface by Cultured Limbal Cells

Mechanisms by which cultured LEC restore the ocular surface are unknown. LEC may (a) replace progenitor/stem cells, and/or (b) revitalize dormant stem cells of the recipient by providing growth factors/chemotactic stimuli. It has been speculated that there may be dormant stem cells despite a clinical picture of total LSCD [39]. Daya et al. [240] suggested that progenitor cells circulating in the bloodstream could populate the ocular surface.

There are several lines of evidence supporting the hypothesis that cultured LEC primarily work by providing a suitable environment for resident cells in the limbus. These include (a) evidence of limited long-term survival of ex vivo expanded LEC on the ocular surface [240], (b) successful treatment of LSCD by alternative cell sources, (c) production of cytokines and growth factors (transforming growth factor α, β1 and β2, platelet-derived growth factor, insulin-like growth factor 1, basic fibroblast growth factor, nerve growth factor, and epidermal growth factor) by epithelial cultures that may stimulate migration and proliferation of undifferentiated cells [256], and (d) postoperative reappearance of palisades of Vogt following transplantation of cultured LEC [122].

4.4.2 Cultured Oral Mucosal Epithelial Cells

Among all therapeutic non-limbal cell types, oral mucosal cells have the advantage of being most extensively documented, with several promising reports describing their transplantation for treating LSCD in animal models [209, 257–259] and humans [210, 257, 260–273]. As most causes of LSCD do not affect oral mucosal cells, these cells may be used for culture and subsequent transplantation. Even in Stevens–Johnson syndrome, an immunological form of LSCD that also affects the oral mucosa, transplantation of cultured oral mucosal epithelial cells has been successful [271]. Oral mucosal cells have greater angiogenic potential [274–276] compared to limbal cells, which may explain some peripheral corneal neovascularization following transplantation [270]. It has been suggested that this neovascularization may regress following anti-angiogenic therapy [235].

4.4.3 Cultured Conjunctival Epithelial Cells

Recently, a human study showed that cultured conjunctival epithelial cells can be used to treat LSCD [150]. This finding supported previous work demonstrating that conjunctival cells in conjunction with limbal cells could successfully treat LSCD [125, 238]. Transplantation of cultured conjunctival epithelial cells has also proven successful in rabbit models of LSCD [212, 227, 277].

4.4.4 Cultured Embryonic Stem Cells

Epithelial progenitor cells induced from cultured ESC were successfully applied as grafts for treating corneal epithelial injury in mice [211]. Pax6-transfected ESC formed a monolayer of epithelium-like cells in vitro and has been applied in a mouse model of LSCD [229].

4.4.5 Cultured Bone Marrow-Derived Mesenchymal Stem Cells

Transplantation of bone marrow-derived MSC was demonstrated as suitable for the treatment of LSCD in rat [213] and rabbit models [278]. MSC induced by corneal stromal cells in a coculture system also proved effective for treating LSCD in rats [279]. In a study by Reinshagen and colleagues [280], however, bone marrow-derived MSC injected under AM grafted onto the cornea did not reverse LSCD.

4.4.6 Cultured Epidermal Epithelial Cells

Epidermal epithelial cells successfully treated LSCD in goats [214, 281].

4.4.7 Cultured Immature Dental Pulp Stem Cells

Transplantation of a tissue-engineered human immature dental pulp stem cell sheet was previously reported to reconstruct the corneal epithelium in a rabbit model of LSCD [215, 228].

4.4.8 Cultured Hair Follicle-Derived Stem Cells

The transplantation of sheets of cultured follicle bulge-derived stem cells led to reconstruction of the damaged ocular surface in mice [216].

4.4.9 Cultured Umbilical Cord Stem Cells

Transplantation of ex vivo expanded umbilical cord lining stem cells was feasible for treating LSCD in rabbits [218].

4.5 Transplantation of Non-Cultured Cells to Treat Limbal Stem Cell Deficiency

4.5.1 Amniotic Membrane

Amniotic membrane (AM) is the innermost layer of the fetal membranes and consists of a single layer of columnar epithelial cells, a thick basement membrane, and an underlying layer of connective tissue [282]. This substrate facilitates epithelialization, demonstrates low or no immunogenicity, and has anti-inflammatory, anti-angiogenic, and anti-scarring properties [283–288]. Amniotic membrane serves as a surrogate niche for LSC [143, 289], although the exact mechanism of action is yet to be proven. Several studies have demonstrated AM transplantation to be useful in reconstructing the corneal surface of patients inflicted with partial LSCD [160, 290–294]. Amniotic membrane transplantation alone is not considered appropriate for treating total LSCD [160]. More recently, in an uncontrolled study, Liang et al. [295] demonstrated that amnion membrane extract is effective for treating ocular chemical burns, especially mild to moderate cases. In combination with AM, transplantation of non-cultured autologous nasal mucosa appears to be an effective method for goblet cell replenishment [217]. This combined procedure may increase the success rate of other methods to treat LSCD [217]. The issue of goblet cell transplantation to increase the success rate of various procedures intended for treatment of LSC deficiency is highly interesting, and more work should be done to elucidate its potential.

4.5.2 Conjunctival Limbal Autograft

In conjunctival limbal autograft (CLAU), limbal tissue attached to a conjunctival carrier is transplanted from the healthy eye of the patient to the contralateral diseased eye [178, 203]. CLAU is performed for unilateral cases of LSCD. A requirement for this procedure is that the donor eye is free of any conditions that may predispose to future LSCD [232, 296, 297]. This restriction significantly limits the use of this technique because many conditions requiring LEC transplants affect both eyes. CLAU has proven successful in treating LSCD in several studies [226, 298–305] and may be combined with AM transplantation [226]. In contrast to previous reports using two large limbal grafts for CLAU, Kheirkhah et al. [306] showed successful reconstruction of a corneal surface with total LSCD by using only one-sixth of the limbal arc length in conjunction with AM transplantation. Autografts have the best long-term outcome, followed by living-related conjunctival limbal allografts (lr-CLAL), whereas cadaver donor allografts have the poorest outcome [237, 301].

4.5.3 Living-Related Conjunctival Limbal Allograft

lr-CLAL is a procedure in which limbal tissue attached to a conjunctival carrier is harvested from a living relative of the patient and transplanted to the diseased eye [178, 301, 307–311]. Systemic immunosuppression seems to be necessary for lr-CLAL, even in the presence of HLA-matched donor tissue [312].

4.5.4 Simple Limbal Epithelial Transplantation

In 2012, Sangwan and coworkers [313] invented a technique, named simple limbal epithelial transplantation (SLET), which proved effective for treating unilateral LSCD in a human trial. The method involves five main steps: (a) the removal of fibrovascular pannus from the diseased cornea, (b) the harvest of a 2×2 mm biopsy of limbal tissue from the contralateral eye, (c) the division of the biopsy into 8–10 pieces, (d) the suturing of AM onto the recipient cornea, and (e) the culture of small pieces of tissue on AM around the center of the cornea. The technique holds great potential as it circumvents the need for cell laboratories, significantly reducing time and costs of procedures, while still using a small biopsy (2×2 mm). Long-term studies including a large number of patients will determine its future role in therapy for LSCD. By substituting AM with a suitable biodegradable synthetic membrane in future studies, this procedure will have the additional advantage of being free of any foreign products.

4.5.5 Keratolimbal Allograft

Keratolimbal allograft (KLAL) is a procedure which involves the transplantation of limbal tissue, together with a small rim of corneal and scleral tissue, harvested from cadaveric eyes [166, 178, 236, 301, 314–316]. This procedure permits the transplantation of large number of stem cells. Patients with total LSCD will require a 360° KLAL, while those with partial LSCD may require only sectoral KLAL. KLAL may be combined with transplantation of AM [166, 317] and lr-CLAL (the Cincinnati procedure) [318].

4.5.6 Homologous Penetrating Central Limbo-Keratoplasty

Homologous penetrating central limbo-keratoplasty (HPCLK) involves the transplantation of eccentrically trephined corneolimbal grafts [208, 247, 255, 319]. This procedure induces a greater immune reaction than conventional penetrating keratoplasty due to the greater antigenicity of limbal tissue [208, 320].

4.6 Postoperative Considerations of Cell-Based Therapy for Limbal Stem Cell Deficiency

Postoperative investigations may include supplementary tests, such as impression cytology [78, 112, 120] and confocal microscopy [78, 122]. Five basic principles are important in postoperative management of any cell-based LSCD therapy. These include (a) immediate control of inflammation, (b) prophylaxis against infections, (c) maintenance of a moist ocular surface, (d) mechanical protection of transplanted tissue, and (e) prevention of immunological rejection (only relevant for allografts). In addition, bevacizumab (Avastin®) injections [99] and sequential sector conjunctival epitheliectomy (SSCE) [123, 182] have proven effective in reversing the reappearance of a fibrovascular pannus after cultured LEC transplantations.

Control of inflammation is achieved by using topical steroids [313] and sometimes systemic steroids [111]. To prevent bacterial infections, antibiotics are applied topically [128] and more rarely, systemically [113] after transplantations. If necessary, topical lubricating eye drops [104, 106, 122, 125, 127, 169] and occasionally occlusion of the tear ducts [127] are used to keep the ocular surface moist. Furthermore, topical hyaluronic acid [80, 108, 124, 127] with its water-retaining properties [321] and topical autologous serum [80, 108, 124, 127, 169, 182], "a tonic for the ailing corneal epithelium" [322], were in some studies applied postoperatively. Mechanical protection of transplanted cultured cells can be achieved in several ways: (a) the placing of a therapeutic contact lens over the graft [81, 106, 109, 111, 115, 116, 123, 124, 126, 127, 171, 182] (e.g., 2 weeks [207], 6 weeks [106], or up to 3 months [111, 116, 126]) after surgery; (b) the use of an AM onlay [127, 169, 182], which will normally dissolve within weeks [223], and is sutured to the surrounding bulbar conjunctiva with [182] or without [169] scleral sutures; (c) induction of ptosis by administering a botulinum toxin injection [80, 169], which will reduce friction of the upper eyelid; and (d) Steri-Strips, which have been used to close the eyelids for 3 [120] or 7 days [113] with [120] or without [113] proceeding subconjunctival injection with steroids [120] (eyelid sutures may replace Steri-Strips [112]).

Limbal non-cultured allografts are at a significantly higher risk of rejection compared with conventional penetrating keratoplasty [323]. Increased susceptibility to rejection is due to greater access to the immune system because of the vascularity of the limbal region [323, 324] and the antigenicity of limbal tissue conferred by a significant number of Langerhans cells [320]. Transplantation of allogeneic tissue necessitates postoperative immunosuppression. Systemic medications used for immunosuppression include cyclosporine [80] (most frequently used) with [121] or without [80] cyclophosphamide and with [121] or without [80] steroids. The optimum dose and duration of immunosuppression following transplantation of allogeneic ex vivo cultured LEC are unknown.

4.7 Other Treatment Approaches for Limbal Stem Cell Deficiency

4.7.1 Conservative Management of Limbal Stem Cell Deficiency

Conservative strategies for the treatment of LCSD include intensive lubrication of the ocular surface [325], bandage contact lenses/scleral lenses [241, 326], and autologous serum eye drops [327–329].

4.7.2 Keratoprosthesis

The use of keratoprostheses dates back to 1853 [330], and since 1960 there have been numerous reports on the use of this therapeutic approach for corneal dysfunction (*see* Chapter 2) [197, 331–340].

4.7.3 Sequential Sector Conjunctival Epitheliectomy

Repeated debridement of migrating conjunctival epithelium in the acute phase following injury is known as SSCE [341]. This procedure can reduce or prevent ingrowth of conjunctival tissue in partial LSCD.

4.7.4 Bevacizumab (Avastin®)

Bevacizumab is a human monoclonal antibody that inhibits vascular endothelial growth factor-induced blood vessel formation. This drug, either in the form of eye drops [342, 343] or subconjunctival injection [344], has been used to treat ocular surface neovascularization. Early, rather than late, treatment with bevacizumab seems to be useful in inhibiting neovascularization and conjunctivalization in LSCD. Bevacizumab injections resulted in simultaneous inhibition of corneal neovascularization and conjunctivalization in a rabbit model of LSCD [345]. Moreover, this medication has recently been shown to revert neovascularization subsequent to transplantation of cultured LEC in humans [99].

4.7.5 Steroids

Steroid pulse therapy (intravenous methylprednisolone for 3–4 days) combined with topical betamethasone at the onset of Stevens–Johnson syndrome and toxic epidermal necrolysis may prevent complications associated with LSCD [346].

4.7.6 Limbal Fibroblast Conditioned Medium

In a mouse model with limbus-to-limbus epithelial debridement, topical treatment with limbal fibroblast conditioned medium resulted in improved corneal phenotype, compared to the control groups treated with skin fibroblast conditioned medium or Dulbecco's serum-free medium [347].

4.7.7 Oxygen

In the acute phase of ocular chemical or thermal burns, oxygen therapy was reported to improve limbal ischemia, accelerate epithelialization, increase corneal transparency, and decrease corneal vascularization [348].

5 Universal International Grading System for Therapy for Limbal Stem Cell Deficiency

Comparisons between studies to treat LSCD are complicated by several factors: (a) the use of a variety of techniques; (b) a heterogeneous population of patients, including many causes of LSCD in different stages; (c) inclusion of both autologous and allogeneic transplants, which may either be living-related with or without matching HLA [311] or cadaveric; (d) miscellaneous criteria for diagnosing LSCD, ranging from clinical assessment to the inclusion of supplementary tests; (e) differences in surgical management and postoperative treatment; and (f) different follow-up period within and between studies.

In 2010, the Cornea Society established an international nomenclature for ocular surface rehabilitative procedures to standardize communication and comparison of techniques and outcomes [349]. There is, however, a pressing need to develop by consensus a universal international grading system for clinical outcomes of LSCD. Objective methods of outcome assessment may include clinical scoring systems [182, 350], such as impression cytology and in vivo microscopy (confocal microscopy [193] and/or spectral domain ocular computer tomography [351]). Ideally, vision should be recorded using eye charts containing symbols with equal spacing between them (LogMAR eye chart), rather than those with symbols that are irregularly spaced apart (Snellen eye chart) [352]. Moreover, visual function questionnaires [353] are important as even a slight improvement in visual acuity may mean a great improvement in quality of life [354].

6 Storage and Transportation of Cultured Limbal Epithelial Cells

Therapy with cultured keratinocytes [355] and LEC [220] has been suggested for more widespread use. In the years to come, this may be true for an increasing number of stem cell technologies. Combining cell laboratories and eye banks is likely to increase the availability of regenerative medicine [356, 357] (Fig. 7). Greater interest in preservation of cultured LEC is evident from the literature [112, 113, 116, 356–367]. Cryopreserved LEC enable subsequent culture of these cells, thus facilitating logistics [119, 368–370].

Furthermore, the more advanced and resource-demanding the technology, the greater the need for centralization [313]. Ahmad et al. [371] suggested in a recent review on LSC therapy that the production of cultured LEC should be centralized for the following reasons: (a) strict regulatory demands make laboratories extremely costly, and (b) requirement of expertise in culturing limbal tissue is pivotal. Given the strict regulations for cell therapy, the number of small units producing tissue for clinical use will abate [372], hence forcing centralization and larger centers, which necessitate effective transportation strategies [371].

Storage technology is important as it allows time for quality control of LEC cultures. Daniels and colleagues [372] proposed that, whenever possible, control measures should be employed to determine the quality of the end product. In 2010, Pellegrini et al. [220] suggested that stringent quality criteria should be adopted for LEC cultures to ensure that they comprise a sufficient number of stem cells to enable long-term epithelial survival. These authors stated that quality control should include a rigorous clonal analysis or the evaluation of the number of cell doublings generated during

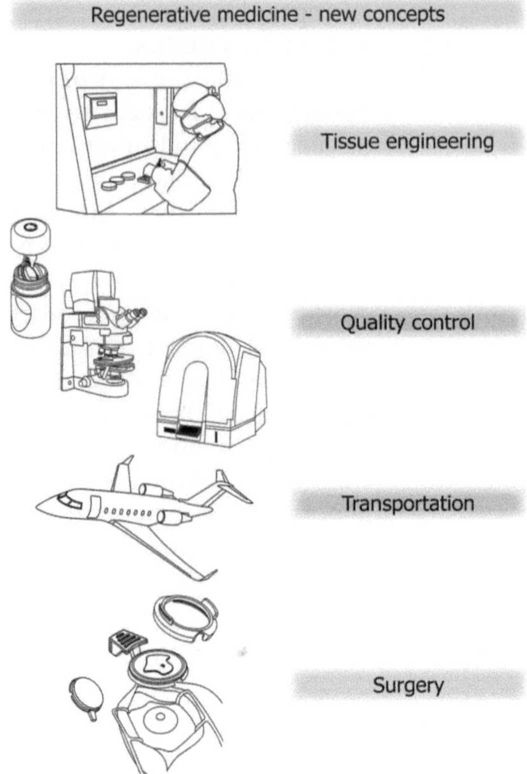

Fig. 7 Storage of cultured cells prepared for quality control and transportation. Courtesy of Håkon Raanes, The Oslo School of Architecture and Design, Norway

serial cultivation of LEC [220]. Regulating levels of contaminating fibroblasts in LEC cultures is another measure for ensuring the quality of these cells. Di Iori and associates [77] monitored the percentage of contamination of murine fibroblasts in cultured LEC, and >5 % contamination was considered unsuitable for clinical use. Others have highlighted the need for standardization [373] and validation [374, 375] of LEC cultures. Proulx et al. [376] stressed the importance of viability assessment of LEC prior to clinical use.

Microbial analyses of LEC cultures are part of the quality control measures conducted before these cells are applied to treat LSCD [377]. Kolli and colleagues [377] reported microbial analyses of medium after 7–10 days of culture. As their total culture period ranged from 12 to 14 days, it is possible that infections could occur at the end of the culture period. Bacterial infections in LEC cultures may be avoided by using a hermetically sealed container for storage prior to clinical use [367]. Tissue storage is also useful to address scheduling issues due to donor variability [378] and production time for cultured cell sheets [379]. Furthermore, as cell cultures may fail at any time during cultivation, the planning of surgery becomes cumbersome without storage technology.

7 Safety Issues in Limbal Stem Cell Deficiency Therapy

To produce and deliver cell therapies in Europe, it is necessary to comply with the European Council (EC) Directives [380–394]. The purpose of EC directives is to allow free movement of products around Europe, and ensure that the best-quality products are used in clinical trials and routine practice to protect patient safety. This involves production in a controlled clean room environment under an extremely stringent quality management system (Good Manufacturing Practice (GMP)) [372]. These demands pose great practical challenges for the cultivation of cells in a dynamic living system compared with drug manufacture.

The use of animal-derived components in cell culture has raised considerable concern [395–397], which has been highlighted most extensively by Schwab et al. [396]. One such component is 3T3 cells (3-day transfer, inoculum 3×10^5 cells), a fibroblast cell line established in 1962 from Swiss mouse embryo tissue [398]. Fetal calf serum (FCS) and 3T3 cells have been used for more than three decades in many countries to culture epithelial cells, including limbal and epidermal cells, to produce transplants for thousands of patients [220]. Coculture systems with 3T3 have been demonstrated to increase colony forming efficiency [399–401] and to decrease the level of differentiation of cultured LEC [402].

Murine 3T3 cells (feeder layers) are treated either by irradiation or with mitomycin C to arrest their growth (prevents engraftment) [396]. Current known methods for growth inactivation of 3T3 cells include irradiation by exposure to cobalt [403] or other gamma sources [396]. The radiation dosage used (60 Gy) is not immediately lethal to cells [404] or antimicrobial [405]. Treatment with mitomycin C is used as an alternative to irradiation, although less is known about cellular lethality or antimicrobial effects of this antitumor agent [396].

There are several possible consequences of the use of 3T3 cells: (a) tumorigenesis [406], (b) xenozoonosis [407, 408], (c) cell fusion [409], (d) xenoantigenicity [410], (e) xenogenic microchimerism [410], and (f) potential contamination with viruses [411] or prion agents [412] during production [396]. Xenozoonosis has not yet been reported with keratinocyte ex vivo expansion, but transmission could be devastating to the recipient as well as the wider community [396]. Due to the inherent risks associated with the use of animal-derived feeder layers, the use of animal-free feeder layers has been proposed [413, 414]. Pellegrini et al. [220], however, put forward that a clone of 3T3 cells has the advantage of producing an identical feeder layer for all cultures, which assures high reproducibility and eliminates the variability of autologous cells. Over the past 30 years, no adverse effects have been reported from the use of clinical grade, GMP-certified 3T3 cells [220]. Furthermore, the International

Society for Stem Cell Research states that "inclusion of animal materials in the cell manufacturing process does not preclude human use," provided appropriate controls and certifications have been made [220].

To further reduce animal-derived components, several investigators have replaced FCS in cell cultures with autologous serum [80, 81, 99, 108, 111, 124, 127, 128, 147, 150, 169, 182, 231, 239, 245]. Although it has been suggested that holoclone formation requires cultures with 3T3 cells and FCS [355, 415, 416], Kolli et al. [182] demonstrated 100 % clinical success rate without 3T3 cells and FCS. Currently, animal-free LEC cultures have been used in 6 reports presenting clinical results [81, 127, 128, 150, 169, 182]. The largest study (200 eyes) on the outcome of transplantation of autologous LEC using an animal-free explant culture technique was recently reported by Sangwan and coworkers [128]. With an average follow-up time of 3 years, 71 % of the patients were successfully treated. The authors, therefore, concluded that this study provided ample clinical evidence to support the continued use of animal-free LEC cultures [128].

8 Future Perspectives

There are several issues to be addressed that may influence the direction of future therapy for LSCD, including the relative importance of cell-based versus non-cell-based therapy. Another interesting issue is whether the future lies in ex vivo or in vivo culture of cells for treating LSCD. For cell-based therapy, the intriguing question as to the cell source and technique that are most suitable in the long term for treating LSCD is still many years away from being answered. Human clinical studies for treating LSCD currently apply limbal, conjunctival, and oral mucosal as no single cell type has proven clearly superior. Comparative long-term, follow-up clinical studies of alternative cell sources together with those already applied in human clinical studies are needed to conclude on the cell type that is able to most effectively reverse LSCD.

It is conceivable that it is not the cells *per se*, but the factors they secrete, that govern the clinical outcome for patients with LSCD. Rather than extending the number of cell types applicable for ocular surface reconstruction, research into the secretome of cultured LEC and their effects may simplify therapy to include eye drops instead of cells. Based on recent advances in therapy for LSCD, it is possible that the future brings a multifaceted approach for treating this disorder, by combining cell-based and non-cell-based therapy.

Acknowledgements

The authors thank Astrid Østerud at the Department of Ophthalmology, Oslo University Hospital, Norway; Ingrid Riphagen at the Unit for Applied Clinical Research, Norwegian University of Science and Technology, Norway; Sten Ræder at the Department of Ophthalmology, Stavanger University Hospital, Stavanger, Norway, and SynsLaser kirurgi, Oslo/Tromsø, Norway; and Øygunn Aass Utheim and Torstein Lyberg at the Department of Medical Biochemistry, Oslo University Hospital, Norway for excellent help and support.

References

1. Weissman BA, Fatt I, Rasson J (1981) Diffusion of oxygen in human corneas in vivo. Invest Ophthalmol Vis Sci 20:123–125
2. Gottsch JD, Chen CH, Aguayo JB, Cousins JP, Strahlman ER, Stark WJ (1986) Glycolytic activity in the human cornea monitored with nuclear magnetic resonance spectroscopy. Arch Ophthalmol 104:886–889
3. Nishida T (2005) Cornea. In: Krachmer JH, Mannis MJ, Holland E (eds) Cornea. Fundamentals, diagnosis and management, 2nd edn. Elsevier, Mosby, Philadelphia, pp 3–26
4. Graymore CN (1970) Biochemistry of the eye. Academic, New York
5. Brandell BW, Goldstick TK, Deutsch TA, Ernest JT (1988) Tissue oxygen uptake from the atmosphere by a new, noninvasive polarographic technique with application to corneal metabolism. Adv Exp Med Biol 222:275–284
6. Rozsa AJ, Beuerman RW (1982) Density and organization of free nerve endings in the corneal epithelium of the rabbit. Pain 14:105–120
7. Beuerman RW, Pedroza L (1996) Ultrastructure of the human cornea. Microsc Res Tech 33:320–335
8. Pfister RR (1973) The normal surface of corneal epithelium: a scanning electron microscopic study. Invest Ophthalmol 12:654–668
9. Klyce SD (1972) Electrical profiles in the corneal epithelium. J Physiol 226:407–429
10. Pajoohesh-Ganji A, Stepp MA (2005) In search of markers for the stem cells of the corneal epithelium. Biol Cell 97:265–276
11. Puangsricharern V, Tseng SC (1995) Cytologic evidence of corneal diseases with limbal stem cell deficiency. Ophthalmology 102:1476–1485
12. Fini ME, Stramer BM (2005) How the cornea heals: cornea-specific repair mechanisms affecting surgical outcomes. Cornea 24:S2–S11
13. Meek KM, Dennis S, Khan S (2003) Changes in the refractive index of the stroma and its extrafibrillar matrix when the cornea swells. Biophys J 85:2205–2212
14. Jester JV, Moller-Pederson T, Huang J, Sax CM, Kays WT, Cavangh HD et al (1999) The cellular basis of corneal transparency: evidence for ‘corneal crystallins’. J Cell Sci 112:613–622
15. Nelson JD, Cameron JD (2005) The conjunctiva: anatomy and physiology. In: Krachmer J, Mannis M, Holland E (eds) Cornea. Fundamentals, diagnosis and management, 2nd edn. Elsevier, Mosby, Philadelphia, pp 37–44
16. Dilly PN (1994) Structure and function of the tear film. Adv Exp Med Biol 350:239–247
17. Tseng SC, Hirst LW, Maumenee AE, Kenyon KR, Sun TT, Green WR (1984) Possible mechanisms for the loss of goblet cells in mucin-deficient disorders. Ophthalmology 91:545–552
18. Wolfley DE (2005) Eyelids. In: Krachmer JH, Mannis MJ, Holland EJ (eds) Cornea. Fundamentals, diagnosis and management, 2nd edn. Elsevier, Mosby, Philadelphia, pp 53–58
19. Doshi S (2004) Corneal anatomy, physiology and response to wounding. In: Naroo SA (ed) Refractive surgery: a guide to assessment and management. BH, Optician, pp 17–21
20. Duke-Elder S (1968) System of ophthalmology. Kimpton, London
21. Davanger M, Evensen A (1971) Role of the pericorneal papillary structure in renewal of corneal epithelium. Nature 229:560–561

22. Cotsarelis G, Cheng SZ, Dong G, Sun TT, Lavker RM (1989) Existence of slow-cycling limbal epithelial basal cells that can be preferentially stimulated to proliferate: implications on epithelial stem cells. Cell 57:201–209
23. Zieske JD (1994) Perpetuation of stem cells in the eye. Eye (Lond) 8:163–169
24. Schermer A, Galvin S, Sun TT (1986) Differentiation-related expression of a major 64 K corneal keratin in vivo and in culture suggests limbal location of corneal epithelial stem cells. J Cell Biol 103:49–62
25. Beebe DC, Masters BR (1996) Cell lineage and the differentiation of corneal epithelial cells. Invest Ophthalmol Vis Sci 37:1815–1825
26. Thoft RA, Friend J (1983) The X, Y, Z hypothesis of corneal epithelial maintenance. Invest Ophthalmol Vis Sci 24:1442–1443
27. Watanabe K, Nakagawa S, Nishida T (1987) Stimulatory effects of fibronectin and EGF on migration of corneal epithelial cells. Invest Ophthalmol Vis Sci 28:205–211
28. Welge-Lussen U, May CA, Neubauer AS, Priglinger S (2001) Role of tissue growth factors in aqueous humor homeostasis. Curr Opin Ophthalmol 12:94–99
29. Rolando M, Zierhut M (2001) The ocular surface and tear film and their dysfunction in dry eye disease. Surv Ophthalmol 45:S203–S210
30. West-Mays JA, Dwivedi DJ (2006) The keratocyte: corneal stromal cell with variable repair phenotypes. Int J Biochem Cell Biol 38:1625–1631
31. Schofield R (1983) The stem cell system. Biomed Pharmacother 37:375–380
32. Watt FM (2000) Out of eden: stem cells and their niches. Science 287:1427–1430
33. Van Buskirk EM (1989) The anatomy of the limbus. Eye (Lond) 3:101–108
34. Gipson IK (1989) The epithelial basement membrane zone of the limbus. Eye (Lond) 3:132–140
35. Schlotzer-Schrehardt U, Dietrich T, Saito K, Sorokin L, Sasaki T, Paulsson M et al (2007) Characterization of extracellular matrix components in the limbal epithelial stem cell compartment. Exp Eye Res 85:845–860
36. Shanmuganathan VA, Foster T, Kulkarni BB, Hopkinson A, Gray T, Powe DG et al (2007) Morphological characteristics of the limbal epithelial crypt. Br J Ophthalmol 91:514–519
37. Dua HS, Shanmuganathan VA, Powell-Richards AO, Tighe PJ, Joseph A (2005) Limbal epithelial crypts: a novel anatomical structure and a putative limbal stem cell niche. Br J Ophthalmol 89:529–532
38. Majo F, Rochat A, Nicolas M, Jaoude GA, Barrandon Y (2008) Oligopotent stem cells are distributed throughout the mammalian ocular surface. Nature 456:250–254
39. Dua HS, Miri A, Alomar T, Yeung AM, Said DG (2009) The role of limbal stem cells in corneal epithelial maintenance: testing the dogma. Ophthalmology 116:856–863
40. Pepose JS, Ubels JL (1992) The cornea. In: Hart WMJ (ed) Adler's physiology of the e ye, 9th edn. Mosby Year Book Inc., St. Louis, pp 29–70
41. Sun TT, Tseng SC, Lavker RM (2010) Location of corneal epithelial stem cells. Nature 463:E10–E11
42. Emamy H, Ahmadian H (1977) Limbal dermoid with ectopic brain tissue. Report of a case and review of the literature. Arch Ophthalmol 95:2201–2202
43. Weinstein JM, Romano PE, O'Grady RB (1979) Bone formation in association with a limbal dermoid. Arch Ophthalmol 97:1121–1122
44. Mann I (1944) A study of epithelial regeneration in the living eye. Br J Ophthalmol 28:26–40
45. Hanna C (1966) Proliferation and migration of epithelial cells during corneal wound repair in the rabbit and the rat. Am J Ophthalmol 61:55–63
46. Dua HS, Forrester JV (1990) The corneoscleral limbus in human corneal epithelial wound healing. Am J Ophthalmol 110:646–656
47. Buck RC (1979) Cell migration in repair of mouse corneal epithelium. Invest Ophthalmol Vis Sci 18:767–784
48. Kinoshita S, Friend J, Thoft RA (1981) Sex chromatin of donor corneal epithelium in rabbits. Invest Ophthalmol Vis Sci 21:434–441
49. Srinivasan BD, Eakins KE (1979) The reepithelialization of rabbit cornea following single and multiple denudation. Exp Eye Res 29:595–600
50. Chen JJ, Tseng SC (1990) Corneal epithelial wound healing in partial limbal deficiency. Invest Ophthalmol Vis Sci 31:1301–1314
51. Chen JJ, Tseng SC (1991) Abnormal corneal epithelial wound healing in partial-thickness removal of limbal epithelium. Invest Ophthalmol Vis Sci 32:2219–2233
52. Huang AJ, Tseng SC (1991) Corneal epithelial wound healing in the absence of limbal epithelium. Invest Ophthalmol Vis Sci 32:96–105

53. Barrandon Y, Green H (1985) Cell size as a determinant of the clone-forming ability of human keratinocytes. Proc Natl Acad Sci U S A 82:5390–5394
54. Romano AC, Espana EM, Yoo SH, Budak MT, Wolosin JM, Tseng SC (2003) Different cell sizes in human limbal and central corneal basal epithelia measured by confocal microscopy and flow cytometry. Invest Ophthalmol Vis Sci 44:5125–5129
55. Chen Z, de Paiva CS, Luo L, Kretzer FL, Pflugfelder SC, Li DQ (2004) Characterization of putative stem cell phenotype in human limbal epithelia. Stem Cells 22:355–366
56. Schlotzer-Schrehardt U, Kruse FE (2005) Identification and characterization of limbal stem cells. Exp Eye Res 81:247–264
57. Watanabe K, Nishida K, Yamato M, Umemoto T, Sumide T, Yamamoto K et al (2004) Human limbal epithelium contains side population cells expressing the ATP-binding cassette transporter ABCG2. FEBS Lett 565:6–10
58. Wolosin JM, Budak MT, Akinci MA (2004) Ocular surface epithelial and stem cell development. Int J Dev Biol 48:981–991
59. Budak MT, Alpdogan OS, Zhou M, Lavker RM, Akinci MA, Wolosin JM (2005) Ocular surface epithelia contain ABCG2-dependent side population cells exhibiting features associated with stem cells. J Cell Sci 118:8–24
60. de Paiva CS, Chen Z, Corrales RM, Pflugfelder SC, Li DQ (2005) ABCG2 transporter identifies a population of clonogenic human limbal epithelial cells. Stem Cells 23:63–73
61. Di Iorio E, Barbaro V, Ruzza A, Ponzin D, Pellegrini G, De Luca M (2005) Isoforms of DeltaNp63 and the migration of ocular limbal cells in human corneal regeneration. Proc Natl Acad Sci U S A 102:9523–9528
62. Chaloin-Dufau C, Sun TT, Dhouailly D (1990) Appearance of the keratin pair K3/K12 during embryonic and adult corneal epithelial differentiation in the chick and in the rabbit. Cell Differ Dev 32:97–108
63. Kiritoshi A, SundarRaj N, Thoft RA (1991) Differentiation in cultured limbal epithelium as defined by keratin expression. Invest Ophthalmol Vis Sci 32:3073–3077
64. Chen WY, Mui MM, Kao WW, Liu CY, Tseng SC (1994) Conjunctival epithelial cells do not transdifferentiate in organotypic cultures: expression of K12 keratin is restricted to corneal epithelium. Curr Eye Res 13:765–778
65. Arpitha P, Prajna NV, Srinivasan M, Muthukkaruppan V (2005) High expression of p63 combined with a large N/C ratio defines a subset of human limbal epithelial cells: implications on epithelial stem cells. Invest Ophthalmol Vis Sci 46:3631–3636
66. Lindberg K, Brown ME, Chaves HV, Kenyon KR, Rheinwald JG (1993) In vitro propagation of human ocular surface epithelial cells for transplantation. Invest Ophthalmol Vis Sci 34:2672–2679
67. Pellegrini G, Golisano O, Paterna P, Lambiase A, Bonini S, Rama P et al (1999) Location and clonal analysis of stem cells and their differentiated progeny in the human ocular surface. J Cell Biol 145:769–782
68. Bickenbach JR (1981) Identification and behavior of label-retaining cells in oral mucosa and skin. J Dent Res 60:1611–1620
69. Lavker RM, Dong G, Cheng SZ, Kudoh K, Cotsarelis G, Sun TT (1991) Relative proliferative rates of limbal and corneal epithelia. Implications of corneal epithelial migration, circadian rhythm, and suprabasally located DNA-synthesizing keratinocytes. Invest Ophthalmol Vis Sci 32:1864–1875
70. Tseng SC, Zhang SH (1995) Limbal epithelium is more resistant to 5-fluorouracil toxicity than corneal epithelium. Cornea 14:394–401
71. Lehrer MS, Sun TT, Lavker RM (1998) Strategies of epithelial repair: modulation of stem cell and transit amplifying cell proliferation. J Cell Sci 111:2867–2875
72. Schwartz GS, Holland EJ (2005) Classification and staging of ocular surface disease. In: Krachmer JH, Mannis MJ, Holland EJ (eds) Cornea, 2nd edn. Elsevier Mosby, Philadelphia, pp 1785–1797
73. Dua HS, Joseph A, Shanmuganathan VA, Jones RE (2003) Stem cell differentiation and the effects of deficiency. Eye (Lond) 17:877–885
74. Vemuganti GK, Sangwan VS (2010) Interview: affordability at the cutting edge: stem cell therapy for ocular surface reconstruction. Reg Anesth 5:337–340
75. Schwab IR, Isseroff RR (2000) Bioengineered corneas—the promise and the challenge. N Engl J Med 343:136–138
76. Espana EM, Grueterich M, Romano AC, Touhami A, Tseng SC (2002) Idiopathic limbal stem cell deficiency. Ophthalmology 109:2004–2010
77. Di Iorio E, Ferrari S, Fasolo A, Bohm E, Ponzin D, Barbaro V (2010) Techniques for culture and assessment of limbal stem cell grafts. Ocul Surf 8:146–153
78. Shortt AJ, Secker GA, Rajan MS, Meligonis G, Dart JK, Tuft SJ et al (2008) Ex vivo expansion and transplantation of limbal epithelial stem cells. Ophthalmology 115:1989–1997

79. Gomes JA, Pazos HS, Silva AB, Cristovam PC, Belfort Junior R (2009) Transplante de celulas-tronco epiteliais limbicas alogenas expandidas ex vivo sobre membrana amniotica: relato de caso. Arq Bras Oftalmol 72:254–256

80. Pauklin M, Fuchsluger TA, Westekemper H, Steuhl KP, Meller D (2010) Midterm results of cultivated autologous and allogeneic limbal epithelial transplantation in limbal stem cell deficiency. Dev Ophthalmol 45:57–70

81. Nakamura T, Inatomi T, Sotozono C, Ang LP, Koizumi N, Yokoi N et al (2006) Transplantation of autologous serum-derived cultivated corneal epithelial equivalents for the treatment of severe ocular surface disease. Ophthalmology 113:1765–1772

82. Nishida K, Kinoshita S, Ohashi Y, Kuwayama Y, Yamamoto S (1995) Ocular surface abnormalities in aniridia. Am J Ophthalmol 120:368–375

83. Desousa JL, Daya S, Malhotra R (2009) Adnexal surgery in patients undergoing ocular surface stem cell transplantation. Ophthalmology 116:235–242

84. Skeens HM, Brooks BP, Holland EJ (2011) Congenital aniridia variant: minimally abnormal irides with severe limbal stem cell deficiency. Ophthalmology 118:1260–1264

85. Kivlin JD, Apple DJ, Olson RJ, Manthey R (1986) Dominantly inherited keratitis. Arch Ophthalmol 104:1621–1623

86. Pearce WG, Mielke BW, Hassard DT, Climenhaga HW, Climenhaga DB, Hodges EJ (1995) Autosomal dominant keratitis: a possible aniridia variant. Can J Ophthalmol 30:131–137

87. Tseng SC, Meller D, Anderson DF, Touhami A, Pires RT, Gruterich M et al (2002) Ex vivo preservation and expansion of human limbal epithelial stem cells on amniotic membrane for treating corneal diseases with total limbal stem cell deficiency. Adv Exp Med Biol 506((Pt B)):1323–1334

88. Tijmes NT, Zaal MJ, De Jong PT, Volker-Dieben HJ (1997) Two families with dyshidrotic ectodermal dysplasia associated with ingrowth of corneal vessels, limbal hair growth, and Bitot-like conjunctival anomalies. Ophthalmic Genet 18:185–192

89. Di Iorio E, Kaye SB, Ponzin D, Barbaro V, Ferrari S, Bohm E et al (2011) Limbal stem cell deficiency and ocular phenotype in ectrodactyly-ectodermal dysplasia-clefting syndrome caused by p63 mutations. Ophthalmology 119:74–83

90. Anderson NJ, Hardten DR, McCarty TM (2003) Penetrating keratoplasty and keratolimbal allograft transplantation for corneal perforations associated with the ectodermal dysplasia syndrome. Cornea 22:385–388

91. Messmer EM, Kenyon KR, Rittinger O, Janecke AR, Kampik A (2005) Ocular manifestations of keratitis-ichthyosis-deafness (KID) syndrome. Ophthalmology 112:e1–e6

92. Gicquel JJ, Lami MC, Catier A, Balayre S, Dighiero P (2002) Insuffisance limbique bilaterale associee a un syndrome KID. A propos d'un cas. J Fr Ophtalmol 25:1061–1064

93. Fernandes M, Sangwan VS, Vemuganti GK (2004) Limbal stem cell deficiency and xeroderma pigmentosum: a case report. Eye (Lond) 18:741–743

94. Inan UU, Yilmaz MD, Demir Y, Degirmenci B, Ermis SS, Ozturk F (2006) Characteristics of lacrimo-auriculo-dento-digital (LADD) syndrome: case report of a family and literature review. Int J Pediatr Otorhinolaryngol 70:1307–1314

95. Cortes M, Lambiase A, Sacchetti M, Aronni S, Bonini S (2005) Limbal stem cell deficiency associated with LADD syndrome. Arch Ophthalmol 123:691–694

96. Traboulsi EI, Azar DT, Jarudi N, Der Kaloustian VM (1985) Ocular findings in the candidiasis-endocrinopathy syndrome. Am J Ophthalmol 99:486–487

97. Tarkkanen A, Merenmies L (2001) Corneal pathology and outcome of keratoplasty in autoimmune polyendocrinopathy-candidiasis-ectodermal dystrophy (APECED). Acta Ophthalmol Scand 79:204–207

98. Usui T, Nakagawa S, Yokoo S, Mimura T, Yamagami S, Amano S (2010) Bilateral limbal stem cell deficiency with chromosomal translocation of 3p and 9p. Jpn J Ophthalmol 54:357–358

99. Thanos M, Pauklin M, Steuhl KP, Meller D (2010) Ocular surface reconstruction with cultivated limbal epithelium in a patient with unilateral stem cell deficiency caused by Epidermolysis Bullosa Dystrophica Hallopeau-Siemens. Cornea 29:462–464

100. Espana EM, Raju VK, Tseng SC (2002) Focal limbal stem cell deficiency corresponding to an iris coloboma. Br J Ophthalmol 86:1451–1452

101. Aslan D, Akata RF (2010) Dyskeratosis congenita and limbal stem cell deficiency. Exp Eye Res 90:472–473

102. Shimazaki J, Shimmura S, Tsubota K (2002) Limbal stem cell transplantation for the treatment of subepithelial amyloidosis of the cornea (gelatinous drop-like dystrophy). Cornea 21:177–180

103. Sahu SK, Das S, Sachdeva V, Sangwan VS (2009) Alcaligenes xylosoxidans keratitis after

autologous cultivated limbal epithelium transplant. Can J Ophthalmol 44:336–337

104. Baradaran-Rafii A, Ebrahimi M, Kanavi MR, Taghi-Abadi E, Aghdami N, Eslani M et al (2010) Midterm outcomes of autologous cultivated limbal stem cell transplantation with or without penetrating keratoplasty. Cornea 29:502–509
105. Colabelli Gisoldi RA, Pocobelli A, Villani CM, Amato D, Pellegrini G (2010) Evaluation of molecular markers in corneal regeneration by means of autologous cultures of limbal cells and keratoplasty. Cornea 29:715–722
106. Fatima A, Vemuganti GK, Iftekhar G, Rao GN, Sangwan VS (2007) In vivo survival and stratification of cultured limbal epithelium. Clin Experiment Ophthalmol 35:96–98
107. Grueterich M (2002) Phenotypic study of a case with successful transplantation of ex vivo expanded human limbal epithelium for unilateral total limbal stem cell deficiency. Ophthalmology 109:1547–1552
108. Kawashima M, Kawakita T, Satake Y, Higa K, Shimazaki J (2007) Phenotypic study after cultivated limbal epithelial transplantation for limbal stem cell deficiency. Arch Ophthalmol 125:1337–1344
109. Koizumi N, Inatomi T, Suzuki T, Sotozono C, Kinoshita S (2001) Cultivated corneal epithelial stem cell transplantation in ocular surface disorders. Ophthalmology 108:1569–1574
110. Nakamura T, Koizumi N, Tsuzuki M, Inoki K, Sano Y, Sotozono C et al (2003) Successful regrafting of cultivated corneal epithelium using amniotic membrane as a carrier in severe ocular surface disease. Cornea 22:70–71
111. Nakamura T, Inatomi T, Sotozono C, Koizumi N, Kinoshita S (2004) Successful primary culture and autologous transplantation of corneal limbal epithelial cells from minimal biopsy for unilateral severe ocular surface disease. Acta Ophthalmol Scand 82:468–471
112. Rama P, Bonini S, Lambiase A, Golisano O, Paterna P, De Luca M et al (2001) Autologous fibrin-cultured limbal stem cells permanently restore the corneal surface of patients with total limbal stem cell deficiency. Transplantation 72:1478–1485
113. Rama P, Matuska S, Paganoni G, Spinelli A, De Luca M, Pellegrini G (2010) Limbal stem-cell therapy and long-term corneal regeneration. N Engl J Med 363:147–155
114. Sangwan VS, Matalia HP, Vemuganti GK, Ifthekar G, Fatima A, Singh S et al (2005) Early results of penetrating keratoplasty after cultivated limbal epithelium transplantation. Arch Ophthalmol 123:334–340
115. Sangwan VS, Matalia HP, Vemuganti GK, Fatima A, Ifthekar G, Singh S et al (2006) Clinical outcome of autologous cultivated limbal epithelium transplantation. Indian J Ophthalmol 54:29–34
116. Schwab IR, Reyes M, Isseroff RR (2000) Successful transplantation of bioengineered tissue replacements in patients with ocular surface disease. Cornea 19:421–426
117. Tsai RJ, Li LM, Chen JK (2000) Reconstruction of damaged corneas by transplantation of autologous limbal epithelial cells. N Engl J Med 343:86–93
118. Pan Z, Zhang W, Wu Y, Sun B (2002) Transplantation of corneal stem cells cultured on amniotic membrane for corneal burn: experimental and clinical study. Chin Med J (Engl) 115:767–769
119. Harkin DG, Barnard Z, Gillies P, Ainscough SL, Apel AJ (2004) Analysis of p63 and cytokeratin expression in a cultivated limbal autograft used in the treatment of limbal stem cell deficiency. Br J Ophthalmol 88:1154–1158
120. Marchini G, Pedrotti E, Pedrotti M, Barbaro V, Di Iorio E, Ferrari S et al (2011) Long-term effectiveness of autologous cultured limbal stem cell grafts in patients with limbal stem cell deficiency due to chemical burns. Clin Exp Ophthalmol 40:255–267
121. Nakamura T, Sotozono C, Bentley AJ, Mano S, Inatomi T, Koizumi N et al (2010) Long-term phenotypic study after allogeneic cultivated corneal limbal epithelial transplantation for severe ocular surface diseases. Ophthalmology 117:2247–2254
122. Sharma S, Tandon R, Mohanty S, Sharma NMV, Sen S, Kashyap S et al (2011) Culture of corneal limbal epithelial stem cells: experience from benchtop to bedside in a tertiary care hospital in India. Cornea 30:1223–1232
123. Dobrowolski D, Wylegala E, Orzechowska-Wylegala B, Wowra B, Wroblewska-Czajka E (2011) Application of autologous cultivated corneal epithelium for corneal limbal stem cell insufficiency—short-term results. Klin Oczna 113:346–351
124. Shigeyasu C, Shimazaki J (2011) Ocular surface reconstruction after exposure to high concentrations of antiseptic solutions. Cornea 31:59–65
125. Sangwan VS, Vemuganti GK, Iftekhar G, Bansal AK, Rao GN (2003) Use of autologous cultured limbal and conjunctival epithelium in a patient with severe bilateral ocular surface disease induced by acid injury: a case report of unique application. Cornea 22:478–481

126. Schwab IR (1999) Cultured corneal epithelia for ocular surface disease. Trans Am Ophthalmol Soc 97:891–986
127. Shimazaki J, Higa K, Morito F, Dogru M, Kawakita T, Satake Y et al (2007) Factors influencing outcomes in cultivated limbal epithelial transplantation for chronic cicatricial ocular surface disorders. Am J Ophthalmol 143:945–953
128. Sangwan VS, Basu S, Vemuganti GK, Sejpal K, Subramaniam SV, Bandyopadhyay S et al (2011) Clinical outcomes of xeno-free autologous cultivated limbal epithelial transplantation: a 10-year study. Br J Ophthalmol 95:1525–1529
129. Kadar T, Horwitz V, Sahar R, Cohen M, Cohen L, Gez R et al (2011) Delayed loss of corneal epithelial stem cells in a chemical injury model associated with limbal stem cell deficiency in rabbits. Curr Eye Res 36:1098–1107
130. Ghasemi H, Ghazanfari T, Ghassemi-Broumand M, Javadi MA, Babaei M, Soroush MR et al (2009) Long-term ocular consequences of sulfur mustard in seriously eye-injured war veterans. Cutan Ocul Toxicol 28:71–77
131. Baradaran-Rafii A, Eslani M, Tseng SC (2011) Sulfur mustard-induced ocular surface disorders. Ocul Surf 9:163–178
132. Kadar T, Dachir S, Cohen L, Sahar R, Fishbine E, Cohen M et al (2009) Ocular injuries following sulfur mustard exposure–pathological mechanism and potential therapy. Toxicology 263:59–69
133. Javadi MA, Jafarinasab MR, Feizi S, Karimian F, Negahban K (2011) Management of mustard gas-induced limbal stem cell deficiency and keratitis. Ophthalmology 1187:1272–1281
134. Pleyer U, Sherif Z, Baatz H, Hartmann C (1999) Delayed mustard gas keratopathy: clinical findings and confocal microscopy. Am J Ophthalmol 128:506–507
135. Javadi MA, Yazdani S, Sajjadi H, Jadidi K, Karimian F, Einollahi B et al (2005) Chronic and delayed-onset mustard gas keratitis: report of 48 patients and review of literature. Ophthalmology 112:617–625
136. Kwok LS, Coroneo MT (1994) A model for pterygium formation. Cornea 13:219–224
137. Mahdy MA, Bhatia J (2009) Treatment of primary pterygium: role of limbal stem cells and conjunctival autograft transplantation. Oman J Ophthalmol 2:23–26
138. Dushku N, John MK, Schultz GS, Reid TW (2001) Pterygia pathogenesis: corneal invasion by matrix metalloproteinase expressing altered limbal epithelial basal cells. Arch Ophthalmol 119:695–706
139. Chui J, Coroneo MT, Tat LT, Crouch R, Wakefield D, Di Girolamo N (2011) Ophthalmic pterygium: a stem cell disorder with premalignant features. Am J Pathol 178:817–827
140. Dua HS, Azuara-Blanco A (1999) Allo-limbal transplantation in patients with limbal stem cell deficiency. Br J Ophthalmol 83:414–419
141. Holland EJ, Schwartz GS (1999) Epithelial stem-cell transplantation for severe ocular-surface disease. N Engl J Med 340:1752–1753
142. Kremer I, Ehrenberg M, Weinberger D (2009) Fresh-tissue corneolimbal covering graft for large corneal perforation following childhood trachoma. Ophthalmic Surg Lasers Imaging 40:245–250
143. Grueterich M, Espana EM, Tseng SC (2003) Ex vivo expansion of limbal epithelial stem cells: amniotic membrane serving as a stem cell niche. Surv Ophthalmol 48:631–646
144. Vemuganti GK, Fatima A, Madhira SL, Basti S, Sangwan VS (2009) Limbal stem cells: Application in ocular biomedicine. Int Rev Cell Mol Biol 275:133–181
145. Sangwan VS, Murthy SI, Vemuganti GK, Bansal AK, Gangopadhyay N, Rao GN (2005) Cultivated corneal epithelial transplantation for severe ocular surface disease in vernal keratoconjunctivitis. Cornea 24:426–430
146. Sangwan VS, Jain V, Vemuganti GK, Murthy SI (2011) Vernal keratoconjunctivitis with limbal stem cell deficiency. Cornea 30:491–496
147. Satake Y, Shimmura S, Shimazaki J. (2009) Cultivated autologous limbal epithelial transplantation for symptomatic bullous keratopathy. BMJ Case Rep pii: bcr11.2008.1239.
148. Tan DT, Ficker LA, Buckley RJ (1996) Limbal transplantation. Ophthalmology 103:29–36
149. Tsai RJ, Tseng SC (1994) Human allograft limbal transplantation for corneal surface reconstruction. Cornea 13:389–400
150. Di Girolamo N, Bosch M, Zamora K, Coroneo MT, Wakefield D, Watson SL (2009) A contact lens-based technique for expansion and transplantation of autologous epithelial progenitors for ocular surface reconstruction. Transplantation 87:1571–1578
151. Hong S, Kim EJ, Seong GJ, Seo KY (2010) Limbal stem cell transplantation for limbal dermoid. Ophthalmic Surg Lasers Imaging 9:1–2
152. Schwartz GS, Holland EJ (2001) Iatrogenic limbal stem cell deficiency: when glaucoma management contributes to corneal disease. J Glaucoma 10:443–445
153. Schwartz GS, Holland EJ (1998) Iatrogenic limbal stem cell deficiency. Cornea 17:31–37

154. Holland EJ, Schwartz GS (1997) Iatrogenic limbal stem cell deficiency. Trans Am Ophthalmol Soc 95:95–107

155. Capella MJ, de Alvarez TJ, de la Paz MF (2011) Insuficiencia limbar secundaria a multiples inyecciones intravitreas. Arch Soc Esp Oftalmol 86:89–92

156. Nghiem-Buffet MH, Gatinel D, Jacquot F, Chaine G, Hoang-Xuan T (2003) Limbal stem cell deficiency following phototherapeutic keratectomy. Cornea 22:482–484

157. Cannon TC, Brown MF, Brown HH (2004) Regarding limbal stem cell deficiency following phototherapeutic keratectomy. Cornea 23:421

158. Fujishima H, Shimazaki J, Tsubota K (1996) Temporary corneal stem cell dysfunction after radiation therapy. Br J Ophthalmol 80:911–914

159. Smith GT, Deutsch GP, Cree IA, Liu CS (2000) Permanent corneal limbal stem cell dysfunction following radiotherapy for orbital lymphoma. Eye (Lond) 14:905–907

160. Pires RT, Chokshi A, Tseng SC (2000) Amniotic membrane transplantation or conjunctival limbal autograft for limbal stem cell deficiency induced by 5-fluorouracil in glaucoma surgeries. Cornea 19:284–287

161. Dudney BW, Malecha MA (2004) Limbal stem cell deficiency following topical mitomycin C treatment of conjunctival-corneal intraepithelial neoplasia. Am J Ophthalmol 137:950–951

162. Lichtinger A, Pe'er J, Frucht-Pery J, Solomon A (2010) Limbal stem cell deficiency after topical mitomycin C therapy for primary acquired melanosis with atypia. Ophthalmology 117:431–437

163. Sauder G, Jonas JB (2006) Limbal stem cell deficiency after subconjunctival mitomycin C injection for trabeculectomy. Am J Ophthalmol 141:1129–1130

164. Ding X, Bishop RJ, Herzlich AA, Patel M, Chan CC (2009) Limbal stem cell deficiency arising from systemic chemotherapy with hydroxycarbamide. Cornea 28:221–223

165. Ellies P, Anderson DF, Topuhami A, Tseng SC (2001) Limbal stem cell deficiency arising from systemic chemotherapy. Br J Ophthalmol 85:373–374

166. Saghizadeh M, Soleymani S, Harounian A, Bhakta B, Troyanovsky SM, Brunken WJ et al (2011) Alterations of epithelial stem cell marker patterns in human diabetic corneas and effects of c-met gene therapy. Mol Vis 17:2177–2190

167. Meller D, Fuchsluger T, Pauklin M, Steuhl KP (2009) Ocular surface reconstruction in graft-versus-host disease with HLA-identical living-related allogeneic cultivated limbal epithelium after hematopoietic stem cell transplantation from the same donor. Cornea 28:233–236

168. Ang LP, Sotozono C, Koizumi N, Suzuki T, Inatomi T, Kinoshita S (2007) A comparison between cultivated and conventional limbal stem cell transplantation for Stevens-Johnson syndrome. Am J Ophthalmol 143:178–180

169. Koizumi N, Inatomi T, Suzuki T, Sotozono C, Kinoshita S (2001) Cultivated corneal epithelial transplantation for ocular surface reconstruction in acute phase of Stevens-Johnson syndrome. Arch Ophthalmol 119:298–300

170. Prabhasawat P, Tseng SC (1997) Impression cytology study of epithelial phenotype of ocular surface reconstructed by preserved human amniotic membrane. Arch Ophthalmol 115:1360–1367

171. Elder MJ, Bernauer W, Leonard J, Dart JK (1996) Progression of disease in ocular cicatricial pemphigoid. Br J Ophthalmol 80:292–296

172. Dua HS, Saini JS, Azuara-Blanco A, Gupta P (2000) Limbal stem cell deficiency: concept, aetiology, clinical presentation, diagnosis and management. Indian J Ophthalmol 48:83–92

173. Mohammadpour M, Javadi MA, Karimian F (2006) Limbal stem cell deficiency in the context of autoimmune polyendocrinopathy. Eur J Ophthalmol 16:870–872

174. Solomon A, Ellies P, Anderson DF, Touhami A, Grueterich M, Espana EM et al (2002) Long-term outcome of keratolimbal allograft with or without penetrating keratoplasty for total limbal stem cell deficiency. Ophthalmology 109:1159–1166

175. Di Girolamo N, Kumar RK, Coroneo MT, Wakefield D (2002) UVB-mediated induction of interleukin-6 and -8 in pterygia and cultured human pterygium epithelial cells. Invest Ophthalmol Vis Sci 43:3430–3437

176. Dua HS, Azuara-Blanco A (2000) Limbal stem cells of the corneal epithelium. Surv Ophthalmol 44:415–425

177. Dua HS, Gomes JA, Singh A (1994) Corneal epithelial wound healing. Br J Ophthalmol 78:401–408

178. Holland EJ (1996) Epithelial transplantation for the management of severe ocular surface disease. Trans Am Ophthalmol Soc 94:677–743

179. Tseng SC (1989) Concept and application of limbal stem cells. Eye (Lond) 3:141–157

180. Kruse FE, Cursiefen C, Seitz B, Volcker HE, Naumann GO, Holbach L (2003) Klassifikation der Erkrankungen der Augenoberflache. Teil I. Ophthalmologe 100:899–915

181. Tseng SC (1996) Regulation and clinical implications of corneal epithelial stem cells. Mol Biol Rep 23:47–58
182. Kolli S, Ahmad S, Lako M, Figueiredo F (2010) Successful clinical implementation of corneal epithelial stem cell therapy for treatment of unilateral limbal stem cell deficiency. Stem Cells 28:597–610
183. Egbert PR, Lauber S, Maurice DM (1977) A simple conjunctival biopsy. Am J Ophthalmol 84:798–801
184. Singh R, Joseph A, Umapathy T, Tint NL, Dua HS (2005) Impression cytology of the ocular surface. Br J Ophthalmol 89:1655–1659
185. Thiel MA, Bossart W, Bernauer W (1997) Improved impression cytology techniques for the immunopathological diagnosis of superficial viral infections. Br J Ophthalmol 81:984–988
186. Tseng SC (1985) Staging of conjunctival squamous metaplasia by impression cytology. Ophthalmology 92:728–733
187. Donisi PM, Rama P, Fasolo A, Ponzin D (2003) Analysis of limbal stem cell deficiency by corneal impression cytology. Cornea 22:533–538
188. Barbaro V, Ferrari S, Fasolo A, Pedrotti E, Marchini G, Sbabo A et al (2010) Evaluation of ocular surface disorders: a new diagnostic tool based on impression cytology and confocal laser scanning microscopy. Br J Ophthalmol 94:926–932
189. Jirsova K, Dudakova L, Kalasova S, Vesela V, Merjava S (2011) The OV-TL 12/30 clone of anti-cytokeratin 7 antibody as a new marker of corneal conjunctivalization in patients with limbal stem cell deficiency. Invest Ophthalmol Vis Sci 52:5892–5898
190. Ramirez-Miranda A, Nakatsu MN, Zarei-Ghanavati S, Nguyen CV, Deng SX (2011) Keratin 13 is a more specific marker of conjunctival epithelium than keratin 19. Mol Vis 17:1652–1661
191. Vera LS, Gueudry J, Delcampe A, Roujeau JC, Brasseur G, Muraine M (2009) In vivo confocal microscopic evaluation of corneal changes in chronic Stevens-Johnson syndrome and toxic epidermal necrolysis. Cornea 28:401–407
192. Kurbanyan K, Sejpal KD, Aldave AJ, Deng SX (2012) In vivo confocal microscopic findings in Lisch corneal dystrophy. Cornea 31:437–441
193. Deng SX, Sejpal KD, Tang Q, Aldave AJ, Lee OL, Yu F (2011) Characterization of limbal stem cell deficiency by in vivo laser scanning confocal microscopy: a microstructural approach. Arch Ophthalmol 130:440–445
194. Zarei-Ghanavati S, Ramirez-Miranda A, Deng SX (2011) Limbal lacuna: a novel limbal structure detected by in vivo laser scanning confocal microscopy. Ophthalmic Surg Lasers Imaging 42:e129–e131
195. Bizheva K, Hutchings N, Sorbara L, Moayed AA, Simpson T (2011) In vivo volumetric imaging of the human corneo-scleral limbus with spectral domain OCT. Biomed Opt Express 2:1794–1802
196. DeRötth A (1940) Plastic repair of conjunctival defects with fetal membrane. Arch Ophthalmol 23:522–525
197. Dohlman CH, Barnes S, Ma J (2005) Keratoprosthesis. In: Krachmer J, Mannis M, Holland E (eds) Cornea. Fundamentals, diagnosis and management, 2nd edn. Elsevier, Mosby, Philadelphia, pp 1719–1728
198. Thoft RA (1979) Conjunctival transplantation as an alternative to keratoplasty. Ophthalmology 86:1084–1092
199. Sorsby A, Symons HM (1946) Amniotic membrane grafts in caustic burns of the eye: burns of the second degree. Br J Ophthalmol 30:337–345
200. Barraquer J (1965) The Cornea World Congress. Butterworths, Washington
201. Thoft RA (1977) Conjunctival transplantation. Arch Ophthalmol 95:1425–1427
202. Thoft RA (1984) Keratoepithelioplasty. Am J Ophthalmol 97:1–6
203. Kenyon KR, Tseng SC (1989) Limbal autograft transplantation for ocular surface disorders. Ophthalmology 96:709–722
204. Turgeon PW, Nauheim RC, Roat MI, Stopak SS, Thoft RA (1990) Indications for keratoepithelioplasty. Arch Ophthalmol 108:233–236
205. Kenyon KR, Rapoza PA (1995) Limbal allograft transplantation for ocular surface disorders. Ophthalmology 102(suppl):102
206. Kwitko S, Marinho D, Barcaro S, Bocaccio F, Rymer S, Fernandes S et al (1995) Allograft conjunctival transplantation for bilateral ocular surface disorders. Ophthalmology 102:1020–1025
207. Pellegrini G, Traverso CE, Franzi AT, Zingirian M, Cancedda M, De Luca M (1997) Long-term restoration of damaged corneal surfaces with autologous cultivated corneal epithelium. Lancet 349:990–993
208. Reinhard T, Sundmacher R, Spelsberg H, Althaus C (1999) Homologous penetrating central limbo-keratoplasty (HPCLK) in bilateral limbal stem cell insufficiency. Acta Ophthalmol Scand 77:663–667
209. Nakamura T, Endo K, Cooper LJ, Fullwood NJ, Tanifuji N, Tsuzuki M et al (2003) The

successful culture and autologous transplantation of rabbit oral mucosal epithelial cells on amniotic membrane. Invest Ophthalmol Vis Sci 44:106–116

210. Nishida K, Yamato M, Hayashida Y, Watanabe K, Yamamoto K, Adachi E et al (2004) Corneal reconstruction with tissue-engineered cell sheets composed of autologous oral mucosal epithelium. N Engl J Med 351:1187–1196

211. Homma R, Yoshikawa H, Takeno M, Kurokawa MS, Masuda C, Takada E et al (2004) Induction of epithelial progenitors in vitro from mouse embryonic stem cells and application for reconstruction of damaged cornea in mice. Invest Ophthalmol Vis Sci 45:4320–4326

212. Tanioka H, Kawasaki S, Yamasaki K, Ang LP, Koizumi N, Nakamura T et al (2006) Establishment of a cultivated human conjunctival epithelium as an alternative tissue source for autologous corneal epithelial transplantation. Invest Ophthalmol Vis Sci 47:3820–3827

213. Ma Y, Xu Y, Xiao Z, Yang W, Zhang C, Song E et al (2006) Reconstruction of chemically burned rat corneal surface by bone marrow-derived human mesenchymal stem cells. Stem Cells 24:315–321

214. Yang X, Qu L, Wang X, Zhao M, Li W, Hua J et al (2007) Plasticity of epidermal adult stem cells derived from adult goat ear skin. Mol Reprod Dev 74:386–396

215. Monteiro BG, Serafim RC, Melo GB, Silva MC, Lizier NF, Maranduba CM et al (2009) Human immature dental pulp stem cells share key characteristic features with limbal stem cells. Cell Prolif 42:587–594

216. Meyer-Blazejewska EA, Call MK, Yamanaka O, Liu H, Schlotzer-Schrehardt U, Kruse FE et al (2010) From hair to cornea: towards the therapeutic use of hair follicle-derived stem cells in the treatment of limbal stem cell deficiency. Stem Cells 29:57–66

217. Chun YS, Park IK, Kim JC (2011) Technique for autologous nasal mucosa transplantation in severe ocular surface disease. Eur J Ophthalmol 21:545–551

218. Reza HM, Ng BY, Gimeno FL, Phan TT, Ang LP (2011) Umbilical cord lining stem cells as a novel and promising source for ocular surface regeneration. Stem Cell Rev 7:935–947

219. Shimazaki J, Shimmura S, Fujishima H, Tsubota K (2000) Association of preoperative tear function with surgical outcome in severe Stevens-Johnson syndrome. Ophthalmology 107:1518–1523

220. Pellegrini G, Rama P, De LM (2010) Vision from the right stem. Trends Mol Med 17:1–7

221. Tsai RJ, Tseng SC (1995) Effect of stromal inflammation on the outcome of limbal transplantation for corneal surface reconstruction. Cornea 14:439–449

222. Holland EJ, Schwartz GS, Nordlund ML (2005) Surgical techniques for ocular surface reconstruction. In: Krachmer JH, Mannis MJ, Holland EJ (eds) Cornea. Fundamentals, diagnosis and management, 2nd edn. Elsevier, Mosby, Philadelphia, pp 1799–1812

223. Shortt AJ, Secker GA, Notara MD, Limb GA, Khaw PT, Tuft SJ et al (2007) Transplantation of ex vivo cultured limbal epithelial stem cells: a review of techniques and clinical results. Surv Ophthalmol 52:483–502

224. Tsubota K, Fukagawa K, Fujihara T, Shimmura S, Saito I, Saito K et al (1999) Regulation of human leukocyte antigen expression in human conjunctival epithelium. Invest Ophthalmol Vis Sci 40:28–34

225. Iwata M, Yagihashi A, Roat MI, Zeevi A, Iwaki Y, Thoft RA (1994) Human leukocyte antigen-class II-positive human corneal epithelial cells activate allogeneic T cells. Invest Ophthalmol Vis Sci 35:3991–4000

226. Santos MS, Gomes JA, Hofling-Lima AL, Rizzo LV, Romano AC, Belfort R Jr (2005) Survival analysis of conjunctival limbal grafts and amniotic membrane transplantation in eyes with total limbal stem cell deficiency. Am J Ophthalmol 140:223–230

227. Ang LP, Tanioka H, Kawasaki S, Ang LP, Yamasaki K, Do TP et al (2010) Cultivated human conjunctival epithelial transplantation for total limbal stem cell deficiency. Invest Ophthalmol Vis Sci 51:758–764

228. Gomes JA, Geraldes Monteiro B, Melo GB, Smith RL, Pereira C, da Silva M, Lizier NF et al (2010) Corneal reconstruction with tissue-engineered cell sheets composed of human immature dental pulp stem cells. Invest Ophthalmol Vis Sci 51:1408–1414

229. Ueno H, Kurokawa MS, Kayama M, Homma R, Kumagai Y, Masuda C et al (2007) Experimental transplantation of corneal epithelium-like cells induced by Pax6 gene transfection of mouse embryonic stem cells. Cornea 26:1220–1227

230. Basu S, Mohamed A, Chaurasia S, Sejpal K, Vemuganti GK, Sangwan VS (2011) Clinical outcomes of penetrating keratoplasty after autologous cultivated limbal epithelial transplantation for ocular surface burns. Am J Ophthalmol 152:917–924

231. Basu S, Ali H, Sangwan VS (2012) Clinical outcomes of repeat autologous cultivated limbal epithelial transplantation for ocular surface burns. Am J Ophthalmol 153:643–650

232. Jenkins C, Tuft S, Liu C, Buckley R (1993) Limbal transplantation in the management of chronic contact-lens-associated epitheliopathy. Eye (Lond) 7:629–633
233. Gillette TE, Chandler JW, Greiner JV (1982) Langerhans cells of the ocular surface. Ophthalmology 89:700–711
234. Kinoshita S, Koizumi N, Sotozono C, Yamada J, Nakamura T, Inatomi T (2004) Concept and clinical application of cultivated epithelial transplantation for ocular surface disorders. Ocul Surf 2:21–33
235. Lim P, Fuchsluger TA, Jurkunas UV (2009) Limbal stem cell deficiency and corneal neovascularization. Semin Ophthalmol 24:139–148
236. Espana EM, Di Pascuale M, Grueterich M, Solomon A, Tseng SC (2004) Keratolimbal allograft in corneal reconstruction. Eye (Lond) 18:406–417
237. Miri A, Al-Deiri B, Dua HS (2010) Long-term outcomes of autolimbal and allolimbal transplants. Ophthalmology 117:1207–1213
238. Sangwan VS, Vemuganti GK, Singh S, Balasubramanian D (2003) Successful reconstruction of damaged ocular outer surface in humans using limbal and conjuctival stem cell culture methods. Biosci Rep 23:169–174
239. Shimazaki J, Aiba M, Goto E, Kato N, Shimmura S, Tsubota K (2002) Transplantation of human limbal epithelium cultivated on amniotic membrane for the treatment of severe ocular surface disorders. Ophthalmology 109:1285–1290
240. Daya SM, Watson A, Sharpe JR, Giledi O, Rowe A, Martin R et al (2005) Outcomes and DNA analysis of ex vivo expanded stem cell allograft for ocular surface reconstruction. Ophthalmology 112:470–477
241. Baylis O, Figueiredo F, Henein C, Lako M, Ahmad S (2011) 13 years of cultured limbal epithelial cell therapy: a review of the outcomes. J Cell Biochem 112:993–1002
242. Sharpe JR, Daya SM, Dimitriadi M, Martin R, James SE (2007) Survival of cultured allogeneic limbal epithelial cells following corneal repair. Tissue Eng 13:123–132
243. Pauklin M, Steuhl KP, Meller D (2009) Characterization of the corneal surface in limbal stem cell deficiency and after transplantation of cultivated limbal epithelium. Ophthalmology 116:1048–1056
244. Pauklin M, Kakkassery V, Steuhl KP, Meller D (2009) Expression of membrane-associated mucins in limbal stem cell deficiency and after transplantation of cultivated limbal epithelium. Curr Eye Res 34:221–230
245. Meller D, Pauklin M, Westekemper H, Steuhl KP (2010) Autologe Transplantation von kultiviertem Limbusepithel. Ophthalmologe 107:1133–1138
246. Fatima A, Matalia HP, Vemuganti GK, Honavar SG, Sangwan VS (2006) Pseudoepitheliomatous hyperplasia mimicking ocular surface squamous neoplasia following cultivated limbal epithelium transplantation. Clin Experiment Ophthalmol 34:889–891
247. Egarth M, Hellkvist J, Claesson M, Hanson C, Stenevi U (2005) Longterm survival of transplanted human corneal epithelial cells and corneal stem cells. Acta Ophthalmol Scand 83:456–461
248. Henderson TR, Findlay I, Mathews PL, Noble BA (2001) Identifying the origin of single corneal cells by DNA fingerprinting: part I—implications for corneal limbal allografting. Cornea 20:400–403
249. Williams KA, Brereton HM, Aggarwal R, Sykes PJ, Turner DR, Russ GR et al (1995) Use of DNA polymorphisms and the polymerase chain reaction to examine the survival of a human limbal stem cell allograft. Am J Ophthalmol 120:342–350
250. Henderson TR, Findlay I, Matthews PL, Noble BA (2001) Identifying the origin of single corneal cells by DNA fingerprinting: part II—application to limbal allografting. Cornea 20:404–407
251. Du Y, Chen J, Funderburgh JL, Zhu X, Li L (2003) Functional reconstruction of rabbit corneal epithelium by human limbal cells cultured on amniotic membrane. Mol Vis 9:635–643
252. Shimazaki J, Kaido M, Shinozaki N, Shimmura S, Munkhbat B, Hagihara M et al (1999) Evidence of long-term survival of donor-derived cells after limbal allograft transplantation. Invest Ophthalmol Vis Sci 40:1664–1668
253. Henderson TR, McCall SH, Taylor GR, Noble BA (1997) Do transplanted corneal limbal stem cells survive in vivo long-term? Possible techniques to detect donor cell survival by polymerase chain reaction with the amelogenin gene and Y-specific probes. Eye (Lond) 11:779–785
254. Djalilian AR, Mahesh SP, Koch CA, Nussenblatt RB, Shen D, Zhuang Z et al (2005) Survival of donor epithelial cells after limbal stem cell transplantation. Invest Ophthalmol Vis Sci 46:803–807
255. Reinhard T, Spelsberg H, Henke L, Kontopoulos T, Enczmann J, Wernet P et al (2004) Long-term results of allogeneic penetrating limbo-keratoplasty in total limbal stem cell deficiency. Ophthalmology 111:775–782

256. Pellegrini G, De Luca M, Arsenijevic Y (2007) Towards therapeutic application of ocular stem cells. Semin Cell Dev Biol 18:805–818

257. Shimazaki J, Higa K, Kato N, Satake Y (2009) Barrier function of cultivated limbal and oral mucosal epithelial cell sheets. Invest Ophthalmol Vis Sci 50:5672–5680

258. Hayashida Y, Nishida K, Yamato M, Watanabe K, Maeda N, Watanabe H et al (2005) Ocular surface reconstruction using autologous rabbit oral mucosal epithelial sheets fabricated ex vivo on a temperature-responsive culture surface. Invest Ophthalmol Vis Sci 46:1632–1639

259. Nakamura T, Kinoshita S (2003) Ocular surface reconstruction using cultivated mucosal epithelial stem cells. Cornea 22:S75–S80

260. Priya CG, Arpitha P, Vaishali S, Prajna NV, Usha K, Sheetal K et al (2011) Adult human buccal epithelial stem cells: identification, ex-vivo expansion, and transplantation for corneal surface reconstruction. Eye (Lond) 25:1641–1649

261. Liu J, Sheha H, Fu Y, Giegengack M, Tseng SC (2011) Oral mucosal graft with amniotic membrane transplantation for total limbal stem cell deficiency. Am J Ophthalmol 152:739–747

262. Nakamura T, Takeda K, Inatomi T, Sotozono C, Kinoshita S (2011) Long-term results of autologous cultivated oral mucosal epithelial transplantation in the scar phase of severe ocular surface disorders. Br J Ophthalmol 95:942–946

263. Kim JH, Chun YS, Lee SH, Mun SK, Jung HS, Lee SH et al (2010) Ocular surface reconstruction with autologous nasal mucosa in cicatricial ocular surface disease. Am J Ophthalmol 149:45–53

264. Chen HC, Chen HL, Lai JY, Chen CC, Tsai YJ, Kuo MT et al (2009) Persistence of transplanted oral mucosal epithelial cells in human cornea. Invest Ophthalmol Vis Sci 50:4660–4668

265. Ma DH, Kuo MT, Tsai YJ, Chen HC, Chen XL, Wang SF et al (2009) Transplantation of cultivated oral mucosal epithelial cells for severe corneal burn. Eye (Lond) 23:1442–1450

266. Satake Y, Dogru M, Yamane GY, Kinoshita S, Tsubota K, Shimazaki J (2008) Barrier function and cytologic features of the ocular surface. Arch Ophthalmol 126:23–28

267. Nakamura T, Inatomi T, Cooper LJ, Rigby H, Fullwood NJ, Kinoshita S (2007) Phenotypic investigation of human eyes with transplanted autologous cultivated oral mucosal epithelial sheets for severe ocular surface diseases. Ophthalmology 114:1080–1088

268. Ang LP, Nakamura T, Inatomi T, Sotozono C, Koizumi N, Yokoi N et al (2006) Autologous serum-derived cultivated oral epithelial transplants for severe ocular surface disease. Arch Ophthalmol 124:1543–1551

269. Inatomi T, Nakamura T, Kojyo M, Koizumi N, Sotozono C, kinoshita S (2006) Ocular surface reconstruction with combination of cultivated autologous oral mucosal epithelial transplantation and penetrating keratoplasty. Am J Ophthalmol 142:757–764

270. Inatomi T, Nakamura T, Koizumi N, Sotozono C, Yokoi N, kinoshita S (2006) Midterm results on ocular surface reconstruction using cultivated autologous oral mucosal epithelial transplantation. Am J Ophthalmol 141:267–275

271. Nakamura T, Inatomi T, Sotozono C, Amemiya T, Kanamura N, Kinoshita S (2004) Transplantation of cultivated autologous oral mucosal epithelial cells in patients with severe ocular surface disorders. Br J Ophthalmol 88:1280–1284

272. Takeda K, Nakamura T, Inatomi T, Sotozono C, Watanabe A, Kinoshita S (2011) Ocular surface reconstruction using the combination of autologous cultivated oral mucosal epithelial transplantation and eyelid surgery for severe ocular surface disease. Am J Ophthalmol 152:195–201

273. Burillon C, Huot L, Justin V, Nataf S, Chapuis F, Decullier E et al (2011) Cultured autologous oral mucosal epithelial cell sheet (CAOMECS) transplantation for the treatment of corneal limbal epithelial stem cell deficiency. Invest Ophthalmol Vis Sci 53:1325–1331

274. Kanayama S, Nishida K, Yamato M, Hayashi R, Sugiyama H, Soma T et al (2007) Analysis of angiogenesis induced by cultured corneal and oral mucosal epithelial cell sheets in vitro. Exp Eye Res 85:772–781

275. Kanayama S, Nishida K, Yamato M, Hayashi R, Maeda N, Okano T et al (2009) Analysis of soluble vascular endothelial growth factor receptor-1 secreted from cultured corneal and oral mucosal epithelial cell sheets in vitro. Br J Ophthalmol 93:263–267

276. Sekiyama E, Nakamura T, Kawasaki S, Sogabe H, Kinoshita S (2006) Different expression of angiogenesis-related factors between human cultivated corneal and oral epithelial sheets. Exp Eye Res 83:741–746

277. Ono K, Yokoo S, Mimura T, Usui T, Miyata K, Araie M et al (2007) Autologous transplantation of conjunctival epithelial cells cultured on amniotic membrane in a rabbit model. Mol Vis 13:1138–1143

278. Omoto M, Miyashita H, Shimmura S, Higa K, Kawakita T, Yoshida S et al (2009) The use of human mesenchymal stem cell-derived feeder cells for the cultivation of transplantable epithelial sheets. Invest Ophthalmol Vis Sci 50:2109–2115
279. Jiang TS, Cai L, Ji WY, Hui YN, Wang YS, Hu D et al (2010) Reconstruction of the corneal epithelium with induced marrow mesenchymal stem cells in rats. Mol Vis 16:1304–1316
280. Reinshagen H, Auw-Haedrich C, Sorg RV, Boehringer D, Eberwein P, Schwartzkopff J et al (2011) Corneal surface reconstruction using adult mesenchymal stem cells in experimental limbal stem cell deficiency in rabbits. Acta Ophthalmol 89:741–748
281. Yang X, Moldovan NI, Zhao Q, Mi S, Zhou Z, Chen D et al (2008) Reconstruction of damaged cornea by autologous transplantation of epidermal adult stem cells. Mol Vis 14:1064–1070
282. van Herendael BJ, Oberti C, Brosens I (1978) Microanatomy of the human amniotic membranes. A light microscopic, transmission, and scanning electron microscopic study. Am J Obstet Gynecol 131:872–880
283. Hao Y, Ma DH, Hwang DG, Kim WS, Zhang F (2000) Identification of antiangiogenic and antiinflammatory proteins in human amniotic membrane. Cornea 19:348–352
284. Ganatra MA (2003) Amniotic membrane in surgery. J Pak Med Assoc 53:29–32
285. Tseng SC, Espana EM, Kawakita T, Di Pascuale MA, Li W, He H et al (2004) How does amniotic membrane work? Ocul Surf 2:177–187
286. Dua HS, Gomes JA, King AJ, Maharajan VS (2004) The amniotic membrane in ophthalmology. Surv Ophthalmol 49:51–77
287. Fernandes M, Sridhar MS, Sangwan VS, Rao GN (2005) Amniotic membrane transplantation for ocular surface reconstruction. Cornea 24:643–653
288. Gomes JA, Romano A, Santos MS, Dua HS (2005) Amniotic membrane use in ophthalmology. Curr Opin Ophthalmol 16:233–240
289. Grueterich M, Espana E, Tseng SC (2002) Connexin 43 expression and proliferation of human limbal epithelium on intact and denuded amniotic membrane. Invest Ophthalmol Vis Sci 43:63–71
290. Tseng, Prabhasawat P, Barton K, Gray T, Meller D (1998) Amniotic membrane transplantation with or without limbal allografts for corneal surface reconstruction in patients with limbal stem cell deficiency. Arch Ophthalmol 116:431–441
291. Anderson DF, Ellies P, Pires RT, Tseng SC (2001) Amniotic membrane transplantation for partial limbal stem cell deficiency. Br J Ophthalmol 85:567–575
292. Gomes JA, Santos MS, Ventura AS, Donato WB, Cunha MC, Hofling-Lima AL (2003) Amniotic membrane with living related corneal limbal/conjunctival allograft for ocular surface reconstruction in Stevens-Johnson syndrome. Arch Ophthalmol 121:1369–1374
293. Sangwan VS, Matalia HP, Vemuganti GK, Rao GN (2004) Amniotic membrane transplantation for reconstruction of corneal epithelial surface in cases of partial limbal stem cell deficiency. Indian J Ophthalmol 52:281–285
294. Kheirkhah A, Casas V, Raju VK, Tseng SC (2008) Sutureless amniotic membrane transplantation for partial limbal stem cell deficiency. Am J Ophthalmol 145:787–794
295. Liang L, Li W, Ling S, Sheha H, Qui W, Li C et al (2009) Amniotic membrane extraction solution for ocular chemical burns. Clin Experiment Ophthalmol 37:855–863
296. Baradaran-Rafii A, Eslani M, Jamali H, Karimian F, Tailor UA, Djalilian AR (2012) Postoperative complications of conjunctival limbal autograft surgery. Cornea. doi:10.1097/ICO.0b013e31823f095d
297. Dua HS, Azuara-Blanco A (2000) Autologous limbal transplantation in patients with unilateral corneal stem cell deficiency. Br J Ophthalmol 84:273–278
298. Kenyon KR (1989) Limbal autograft transplantation for chemical and thermal burns. Dev Ophthalmol 18:53–58
299. Tsai RJ, Sun TT, Tseng SC (1990) Comparison of limbal and conjunctival autograft transplantation in corneal surface reconstruction in rabbits. Ophthalmology 97:446–455
300. Basti S, Rao SK (2000) Current status of limbal conjunctival autograft. Curr Opin Ophthalmol 11:224–232
301. Wylegala E, Drobrowolski D, Tarnawska D, Janiszewska D, Gabryel B, Malecki A et al (2008) Limbal stem cells transplantation in the reconstruction of the ocular surface: 6 years experience. Eur J Ophthalmol 18:886–890
302. Ronk JF, Ruiz-Esmenjaud S, Osorio M, Bacigalupi M, Goosey JD (1994) Limbal conjunctival autograft in a subacute alkaline corneal burn. Cornea 13:465–468
303. Rao SK, Rajagopal R, Sitalakshmi G, Padmanabhan P (1999) Limbal autografting: comparison of results in the acute and chronic phases of ocular surface burns. Cornea 18:164–171

304. Meallet MA, Espana EM, Grueterich M, Ti SE, Goto E, Tseng SC (2003) Amniotic membrane transplantation with conjunctival limbal autograft for total limbal stem cell deficiency. Ophthalmology 110:1585–1592
305. Yao YF, Zhang B, Zhou P, Jiang JK (2002) Autologous limbal grafting combined with deep lamellar keratoplasty in unilateral eye with severe chemical or thermal burn at late stage. Ophthalmology 109:2011–2017
306. Kheirkhah A, Raju VK, Tseng SC (2008) Minimal conjunctival limbal autograft for total limbal stem cell deficiency. Cornea 27:730–733
307. Huang T, Wang Y, Zhang H, Gao N, Hu A (2011) Limbal allografting from living-related donors to treat partial limbal deficiency secondary to ocular chemical burns. Arch Ophthalmol 129:1267–1273
308. Javadi MA, Baradaran-Rafii A (2009) Living-related conjunctival-limbal allograft for chronic or delayed-onset mustard gas keratopathy. Cornea 28:51–57
309. Tsubota K, Shimmura S, Shinozaki N, Holland EJ, Shimazaki J (2002) Clinical application of living-related conjunctival-limbal allograft. Am J Ophthalmol 133:134–135
310. Daya SM, Ilari FA (2001) Living related conjunctival limbal allograft for the treatment of stem cell deficiency. Ophthalmology 108:126–134
311. Rao SK, Rajagopal R, Sitalakshmi G, Padmanabhan P (1999) Limbal allografting from related live donors for corneal surface reconstruction. Ophthalmology 106:822–828
312. Ozdemir O, Tekeli O, Ornek K, Arslanpence A, Yalcindag NF (2004) Limbal autograft and allograft transplantations in patients with corneal burns. Eye (Lond) 18:241–248
313. Sangwan VS, Basu S, Macneil S, Balasubramanian D (2012) Simple limbal epithelial transplantation (SLET): a novel surgical technique for the treatment of unilateral limbal stem cell deficiency. Br J Ophthalmol. doi:10.1136/bjophthalmol-2011-301164
314. Ilari L, Daya SM (2002) Long-term outcomes of keratolimbal allograft for the treatment of severe ocular surface disorders. Ophthalmology 109:1278–1284
315. Holland EJ, Djalilian AR, Schwartz GS (2003) Management of aniridic keratopathy with keratolimbal allograft: a limbal stem cell transplantation technique. Ophthalmology 110:125–130
316. Liang L, Sheha H, Tseng SC (2009) Long-term outcomes of keratolimbal allograft for total limbal stem cell deficiency using combined immunosuppressive agents and correction of ocular surface deficits. Arch Ophthalmol 127:1428–1434
317. Espana EM, Grueterich M, Ti SE, Tseng SC (2003) Phenotypic study of a case receiving a keratolimbal allograft and amniotic membrane for total limbal stem cell deficiency. Ophthalmology 110:481–486
318. Biber JM, Skeens HM, Neff KD, Holland EJ (2011) The cincinnati procedure: technique and outcomes of combined living-related conjunctival limbal allografts and keratolimbal allografts in severe ocular surface failure. Cornea 30:765–771
319. Spelsberg H, Reinhard T, Henke L, Berschick P, Sundmacher R (2004) Penetrating limbo-keratoplasty for granular and lattice corneal dystrophy: survival of donor limbal stem cells and intermediate-term clinical results. Ophthalmology 111:1528–1533
320. Niederkorn JY (1995) Effect of cytokine-induced migration of Langerhans cells on corneal allograft survival. Eye (Lond) 9:215–218
321. Rah MJ (2011) A review of hyaluronan and its ophthalmic applications. Optometry 82:38–43
322. Pflugfelder SC (2006) Is autologous serum a tonic for the ailing corneal epithelium? Am J Ophthalmol 142:316–317
323. Djalilian AR, Nussenblatt RB, Holland HJ (2001) Immunosuppressive therapy in ocular surface transplantation. In: Holland HJ, Mannis MJ (eds) Ocular surface disease: medical and surgical management. Springer, New York, pp 243–252
324. Goldberg MF, Bron AJ (1982) Limbal palisades of Vogt. Trans Am Ophthalmol Soc 80:155–171
325. Jeng BH, Halfpenny CP, Maisler DM, Stock EL (2010) Management of focal limbal stem cell deficiency associated with soft contact lens wear. Cornea 30:18–23
326. Schornack MM (2011) Limbal stem cell disease: management with scleral lenses. Clin Exp Optom 94:592–594
327. Poon AC, Geerling G, Dart JK, Fraenkel GE, Daniels JT (2001) Autologous serum eyedrops for dry eyes and epithelial defects: clinical and in vitro toxicity studies. Br J Ophthalmol 85:1188–1197
328. Geerling G, Maclennan S, Hartwig D (2004) Autologous serum eye drops for ocular surface disorders. Br J Ophthalmol 88:1467–1474
329. Young AL, Cheng AC, Ng HK, Cheng LL, Leung GY, Lam DS (2004) The use of autologous serum tears in persistent corneal epithelial defects. Eye (Lond) 18:609–614

330. Nussbaum JN (1853) Cornea artificialis, ein Substitut für die Transplantatio corneae. Dtsch Klin ed A Göschen 5:367–378

331. Sejpal K, Yu F, Aldave AJ (2011) The Boston keratoprosthesis in the management of corneal limbal stem cell deficiency. Cornea 30:1187–1194

332. Colby KA, Koo EB (2011) Expanding indications for the Boston keratoprosthesis. Curr Opin Ophthalmol 22:267–273

333. Gomaa A, Comyn O, Liu C (2010) Keratoprostheses in clinical practice—a review. Clin Experiment Ophthalmol 38:211–224

334. Tan XW, Perera AP, Tan A, Tan D, Khor KA, Beuerman RW et al (2011) Comparison of candidate materials for a synthetic osteo-odonto keratoprosthesis device. Invest Ophthalmol Vis Sci 52:21–29

335. Tan DT, Tay AB, Theng JT, Lye KW, Parthasarathy A, Por YM et al (2008) Keratoprosthesis surgery for end-stage corneal blindness in asian eyes. Ophthalmology 115:503–510

336. Tan A, Tan DT, Tan XW, Mehta JS (2012) Osteo-odonto keratoprosthesis: systematic review of surgical outcomes and complication rates. Ocul Surf 10:15–25

337. Michael R, Charoenrook V, de la Paz MF, Hitzl W, Temprano J, Barraquer RI (2008) Long-term functional and anatomical results of osteo- and osteoodonto-keratoprosthesis. Graefes Arch Clin Exp Ophthalmol 246:1133–1137

338. Liu C, Hille K, Tan D, Hicks C, Herold J (2008) Keratoprosthesis surgery. Dev Ophthalmol 41:171–186

339. de la Paz MF, De Toledo JA, Charoenrook V, Sel S, Temprano J, Barraquer RI et al (2011) Impact of clinical factors on the long-term functional and anatomic outcomes of osteo-odonto-keratoprosthesis and tibial bone keratoprosthesis. Am J Ophthalmol 151:829–839

340. de Araujo AL, Charoenrook V, de la Paz MF, Temprano J, Barraquer RI, Michael R (2011) The role of visual evoked potential and electroretinography in the preoperative assessment of osteo-keratoprosthesis or osteo-odonto-keratoprosthesis surgery. Acta Ophthalmol. doi:10.1111/j.1755-3768.2010.02086.x

341. Dua HS (1998) The conjunctiva in corneal epithelial wound healing. Br J Ophthalmol 82:1407–1411

342. Wu PC, Kuo HK, Tai MH, Shin SJ (2009) Topical bevacizumab eyedrops for limbal-conjunctival neovascularization in impending recurrent pterygium. Cornea 28:103–104

343. Bock F, Konig Y, Kruse F, Baier M, Cursiefen C (2008) Bevacizumab (Avastin) eye drops inhibit corneal neovascularization. Graefes Arch Clin Exp Ophthalmol 246:281–284

344. Chen WL, Lin CT, Lin NT, Tu IH, Li JW, Chow LP et al (2009) Subconjunctival injection of bevacizumab (avastin) on corneal neovascularization in different rabbit models of corneal angiogenesis. Invest Ophthalmol Vis Sci 50:1659–1665

345. Lin CT, Hu FR, Kuo KT, Chen YM, Chu HS, Lin YH et al (2010) The different effects of early and late bevacizumab (Avastin) injection on inhibiting corneal neovascularization and conjunctivalization in rabbit limbal insufficiency. Invest Ophthalmol Vis Sci 51:6277–6285

346. Araki Y, Sotozono C, Inatomi T, Ueta M, Yokoi N, Ueda E et al (2009) Successful treatment of Stevens-Johnson syndrome with steroid pulse therapy at disease onset. Am J Ophthalmol 147:1004–1011, 1011

347. Amirjamshidi H, Milani BY, Sagha HM, Movahedan A, Shafiq MA, Lavker RM et al (2011) Limbal fibroblast conditioned media: a non-invasive treatment for limbal stem cell deficiency. Mol Vis 17:658–666

348. Sharifipour F, Baradaran-Rafii A, Idani E, Zamani M, Jabbarpoor Bonyadi MH (2011) Oxygen therapy for acute ocular chemical or thermal burns: a pilot study. Am J Ophthalmol 151:823–828

349. Daya SM, Chan CC, Holland EJ (2011) Cornea Society nomenclature for ocular surface rehabilitative procedures. Cornea 30:1115–1119

350. Schwartz GS, Gomes JA, Holland EJ (2002) Preoperative staging of disease severity. In: Holland EJ, Mannis MJ (eds) Ocular surface disease: medical and surgical management. Springer, New York, pp 158–160

351. Lathrop KL, Gupta D, Kagemann L, Schuman JS, Sundarraj N (2012) Optical coherence tomography as a rapid, accurate, non-contact method of visualizing the palisades of Vogt. Invest Ophthalmol Vis Sci 53:1381–1387

352. Flom MC, Weymouth W, Kahnemad D (1963) Visual resolution and contour interaction. J Opt Soc Am 53:1026–1032

353. Mangione CM, Lee PP, Gutierrez PR, Spritzer K, Berry S, Hays RD et al (2001) Development of the 25-item National eye institute visual function questionnaire. Arch Ophthalmol 119:1050–1058

354. Miri A, Mathew M, Dua HS (2010) Quality of life after limbal transplants. Ophthalmology 117:638

355. De Luca M, Pellegrini G, Green H (2006) Regeneration of squamous epithelia from stem cells of cultured grafts. Regen Med 1:45–57
356. Tandon R et al. (2007) Experience with using an outsourcing model for limbal stem cell culture and transplantation in a tropical developing country. Invest Ophthalmol Vis Sci. 48, E-Abstract 473. www.iovs.org
357. Sotozono C et al. (2009) Multicenter prospective analysis of cultured corneal epithelial sheet transplantation. Invest Ophthalmol Vis Sci. 50, E-Abstract 1519. www.iovs.org
358. Kito K, Kagami H, Kobayashi C, Ueda M, Terasaki H (2005) Effects of cryopreservation on histology and viability of cultured corneal epithelial cell sheets in rabbit. Cornea 24:735–741
359. Utheim TP, Raeder S, Utheim OA, Cai Y, Roald B, Drolsum L et al (2007) A novel method for preserving cultured limbal epithelial cells. Br J Ophthalmol 91:797–800
360. Raeder S, Utheim TP, Utheim OA, Nicolaissen B, Roald B, Cai Y et al (2007) Effects of organ culture and optisol-GS storage on structural integrity, phenotypes, and apoptosis in cultured corneal epithelium. Invest Ophthalmol Vis Sci 48:5484–5493
361. Yeh HJ, Yao CL, Chen HI, Cheng HC, Hwang SM (2008) Cryopreservation of human limbal stem cells ex vivo expanded on amniotic membrane. Cornea 27:327–333
362. Totey, S. et al. (2005) Tissue system with undifferentiated stem cells derived from corneal limbus. Appl.nr. PCT/IB2005/000203, Patent nr. WO/2005/079145. India
363. Raeder, S., et al. (2009) Genome-wide transcriptional analysis of cultured human limbal epithelial cells following hypothermic storage Optisol-GS. Invest Ophthalmol. Vis. Sci. 50, E-Abstract 1773/A441. www.iovs.org
364. Oh JY, Kim MK, Shin KS, Shin MS, Wee WR, Lee JH et al (2007) Efficient cryopreservative conditions for cultivated limbal and conjunctival epithelial cells. Cornea 26:840–846
365. Raeder S, Utheim TP, Messelt E, Lyberg T (2010) The impact of de-epithelialization of the amniotic membrane matrix on morphology of cultured human limbal epithelial cells subject to eye bank storage. Cornea 29:439–445
366. Utheim TP et al. (2011) Comparison of storage of cultured human limbal epithelial cells in HEPES-MEM, Optisol-GS and PAA Quantum 286 for 4 days at 23 °C. Invest Ophthalmol Vis Sci. 52, E-Abstract 5124. www.iovs.org.
367. Utheim TP, Raeder S, Utheim OA, de la Paz M, Raold B, Lyberg T (2009) Sterility control and long-term eye-bank storage of cultured human limbal epithelial cells for transplantation. Br J Ophthalmol 93:980–983
368. Mi S, Yang X, Zhao Q, Qu L, Chen SM, Meek K et al (2008) Reconstruction of corneal epithelium with cryopreserved corneal limbal stem cells in a goat model. Mol Reprod Dev 75:1607–1616
369. Bratanov M, Neronov A, Nikolova E (2009) Limbal explants from cryopreserved cadaver human corneas. Immunofluorescence and light microscopy of epithelial cells growing in culture. Cryo Letters 30:183–189
370. Qu L, Yang X, Wang X, Zhao M, Mi S, Dou Z et al (2009) Reconstruction of corneal epithelium with cryopreserved corneal limbal stem cells in a rabbit model. Vet J 179:392–400
371. Ahmad S, Osei-Bempong C, Dana R, Jurkunas U (2010) The culture and transplantation of human limbal stem cells. J Cell Physiol 225:15–19
372. Daniels JT, Secker GA, Shortt AJ, Tuft SJ, Seetharaman S (2006) Stem cell therapy delivery: treading the regulatory tightrope. Regen Med 1:715–719
373. Zakaria N, Koppen C, Van Tendeloo V, Berneman Z, Hopkinson A, Tassignon MJ (2010) Standardized limbal epithelial stem cell graft generation and transplantation. Tissue Eng Part C Methods 16:921–927
374. Hayashi R, Yamato M, Takayanagi H, Oie Y, Kubota A, Hori Y et al (2010) Validation system of tissue engineered epithelial cell sheets for corneal regenerative medicine. Tissue Eng Part C Methods 16:553–560
375. Higa K, Shimazaki J (2008) Recent advances in cultivated epithelial transplantation. Cornea 27:S41–S47
376. Proulx S, Fradette J, Gauvin R, Larouche D, Germain L (2010) Stem cells of the skin and cornea: their clinical applications in regenerative medicine. Curr Opin Organ Transplant. doi:10.1097/MOT.0b013e32834254f1
377. Kolli S, Lako M, Figueiredo F, Mudhar H, Ahmad S (2008) Loss of corneal epithelial stem cell properties in outgrowths from human limbal explants cultured on intact amniotic membrane. Regen Med 3:329–342
378. Utheim TP, Raeder S, Olstad OK, Utheim OA, de la Paz M, Cheng R et al (2009) Comparison of the histology, gene expression profile, and phenotype of cultured human limbal epithelial cells from different limbal regions. Invest Ophthalmol Vis Sci 50:5165–5172

379. O'Callaghan AR, Daniels JT (2011) Limbal epithelial stem cell therapy: controversies and challenges. Stem Cells 29:1923–1932
380. (2011) Directive 2011/62/EU of the European Parliament and of the Council of 8 June 2011 amending Directive 2001/83/EC on the Community code relating to medicinal products for human use, as regards the prevention of the entry into the legal supply chain of falsified medicinal products (Text with EEA relevance)
381. (2010) Directive 2010/84/EU of the European Parliament and of the Council of 15 December 2010 amending, as regards pharmacovigilance, Directive 2001/83/EC on the Community code relating to medicinal products for human use (Text with EEA relevance)
382. (2009) Directive 2009/33/EC of the European Parliament and of the Council of 18 June 2009 amending Directive 2001/82/EC and Directive 2001/83/EC, as regards variations to the terms of marketing authorisations for medicinal products (Text with EEA relevance)
383. (2008) Directive 2008/29/EC of the European Parliament and of the Council of 11 March 2008 amending Directive 2001/83/EC on the Community code relating to medicinal products for human use, as regards the implementation powers conferred on the Commission
384. (2007) Regulation (EC) No 1394/2007 of the European Parliament and of the Council of 13 November 2007 on advanced therapy medicinal products and amending Directive 2001/83/EC and Regulation (EC) No 726/2004 (Text with EEA relevance)
385. (2006) Regulation (EC) No 1901/2006 of the European Parliament and of the Council of 12 December 2006 on medicinal products for paediatric use and amending Regulation (EEC) No 1768/92, Directive 2001/20/EC, Directive 2001/83/EC and Regulation (EC) No 726/2004 (Text with EEA relevance)
386. (2004) Directive 2004/27/EC of the European Parliament and of the Council of 31 March 2004 amending Directive 2001/83/EC on the Community code relating to medicinal products for human use (Text with EEA relevance)
387. (2004) Directive 2004/24/EC of the European Parliament and of the Council of 31 March 2004 amending, as regards traditional herbal medicinal products, Directive 2001/83/EC on the Community code relating to medicinal products for human use
388. (2003) Commission Directive 2006/63/EC of 25 June 2003 amending Directive 2001/83/EC of the European Parliament and of the Council on the Community code relating to medicinal products of human use (Text with EEA relevance)
389. (2002) Directive 2001/83/EC of the European Parliament and of the Council of 6 November 2001 on the Community code relating to medicinal products for human use. 2001
390. (2001) Directive 2001/83/EC of the European Parliament and of the Council of 6 November 2001 on the Community code relating to medicinal products for human use
391. (2003) Commission Directive 2003/94/EC, of 8 October 2003, laying down the principles and guidelines of good manufacturing practice in respect of medicinal products for human use and investigational medicinal products for human use
392. (2004) Directive 2004/23/EC of the European Parliament and of the Council of 31 March 2004 on setting standards of quality and safety for the donation, procurement, testing, processing, preservation, storage and distribution of human tissues and cells
393. (2006) Commission Directive 2006/17/EC of 8 February 2006 implementing Directive 2004/23/EC of the European Parliament and of the Council as regards certain technical requirements for the donation, procurement and testing of human tissues and cells
394. (2006) Commission Directive 2006/86/EC of 24 October 2006 implementing Directive 2004/23/EC of the European Parliament and of the Council as regards traceability requirements, notification of serious adverse reactions and events and certain technical requirements for the coding, processing, preservation, storage and distribution of human tissues and cells
395. Nakamura T, Kinoshita S (2011) New hopes and strategies for the treatment of severe ocular surface disease. Curr Opin Ophthalmol 22:274–278
396. Schwab IR, Johnson NT, Harkin DG (2006) Inherent risks associated with manufacture of bioengineered ocular surface tissue. Arch Ophthalmol 124:1734–1740
397. Mason SL, Stewart RM, Kearns VR, Williams RL, Sheridan CM (2011) Ocular epithelial transplantation: current uses and future potential. Regen Med 6:767–782
398. Todaro GJ, Green H (1963) Quantitative studies of the growth of mouse embryo cells in culture and their development into established lines. J Cell Biol 17:299–313
399. Tseng SC, Kruse FE, Merritt J, Li DQ (1996) Comparison between serum-free and fibroblast-cocultured single-cell clonal culture systems:

evidence showing that epithelial anti-apoptotic activity is present in 3 T3 fibroblast-conditioned media. Curr Eye Res 15:973–984

400. Cristovam PC, Gloria MA, Melo GB, Gomes JA (2008) Importancia do co-cultivo com fibroblastos de camundongo 3 T3 para estabelecer cultura de suspensao de celulas epiteliais do limbo humano. Arq Bras Oftalmol 71:689–694

401. Chen SY, Hayashida Y, Chen MY, Xie HT, Tseng SC (2010) A new isolation method of human limbal progenitor cells by maintaining close association with their niche cells. Tissue Eng Part C Methods 17:537–548

402. Grueterich M, Espana EM, Tseng SC (2003) Modulation of keratin and connexin expression in limbal epithelium expanded on denuded amniotic membrane with and without a 3 T3 fibroblast feeder layer. Invest Ophthalmol Vis Sci 44:4230–4236

403. Aiuti F, Ensoli F, Fiorelli V, Mezzaroma I, Pinter E, Guerra E et al (1993) Silent HIV infection. Vaccine 11:538–541

404. Tolmach LJ (1961) Growth patterns in x-irradiated HeLa cells. Ann N Y Acad Sci 95:743–757

405. Pruss A, Kao M, Gohs U, Koscielny J, von Versen R, Pauli G (2002) Effect of gamma irradiation on human cortical bone transplants contaminated with enveloped and non-enveloped viruses. Biologicals 30:125–133

406. Rubin H (2005) Degrees and kinds of selection in spontaneous neoplastic transformation: an operational analysis. Proc Natl Acad Sci U S A 102:9276–9281

407. Brewer LA, Lwamba HC, Murtaugh MP, Palmenberg AC, Brown C, Njenga MK (2001) Porcine encephalomyocarditis virus persists in pig myocardium and infects human myocardial cells. J Virol 75:11621–11629

408. Meng XJ, Purcell RH, Halbur PG, Lehman JR, Webb DM, Tsareva TS et al (1997) A novel virus in swine is closely related to the human hepatitis E virus. Proc Natl Acad Sci U S A 94:9860–9865

409. Vassilopoulos G, Russell DW (2003) Cell fusion: an alternative to stem cell plasticity and its therapeutic implications. Curr Opin Genet Dev 13:480–485

410. Hultman CS, Brinson GM, Siltharm S, deSerres S, Cairns BA, Peterson HD et al (1996) Allogeneic fibroblasts used to grow cultured epidermal autografts persist in vivo and sensitize the graft recipient for accelerated second-set rejection. J Trauma 41:51–58

411. Ramarli D, Giri A, Reina S, Poffe O, Cancedda R, Varnier O et al (1995) HIV-1 spreads from lymphocytes to normal human keratinocytes suitable for autologous and allogenic transplantation. J Invest Dermatol 105:644–647

412. Boneva RS, Folks TM, Chapman LE (2001) Infectious disease issues in xenotransplantation. Clin Microbiol Rev 14:1–14

413. Notara M, Haddow DB, MacNeil S, Daniels JT (2007) A xenobiotic-free culture system for human limbal epithelial stem cells. Regen Med 2:919–927

414. Sharma SM, Fuchsluger T, Ahmad S, Katikireddy KR, Armant M, Dana R et al (2011) Comparative analysis of human-derived feeder layers with 3 T3 fibroblasts for the ex vivo expansion of human limbal and oral epithelium. Stem Cell Rev. doi:10.1007/s12015-011-9319-6

415. Rheinwald JG, Green H (1975) Serial cultivation of strains of human epidermal keratinocytes: the formation of keratinizing colonies from single cells. Cell 6:331–343

416. Green H, Kehinde O, Thomas J (1979) Growth of cultured human epidermal cells into multiple epithelia suitable for grafting. Proc Natl Acad Sci U S A 76:5665–5668

Chapter 2

The Artificial Cornea

Naresh Polisetti, Mohammad Mirazul Islam, and May Griffith

Abstract

Human corneal transplantation to date suffers from the shortage of good-quality donor tissue, and in some conditions, allografting is contraindicated. A range of artificial replacements to donor allograft corneas have been developed. These range from keratoprostheses (KPro) that replace basic corneal functions of light transmission and protection to regenerative medicine strategies for regenerating one or more layers of the human cornea. This chapter reviews the advances made in developing artificial corneas or more accurately, artificial alternatives to donor allograft corneas for ocular application.

Key words Artificial cornea, Collagen hydrogel, Corneal blindness, Corneal replacement, Corneal regeneration

1 Introduction

The cornea is the transparent front covering of the eye. It serves as a protective barrier against pathogenic attack and is also the main refractive part of the eye that is responsible for 65–75 % of light transmission to the retina, to enable vision [1]. It comprises three cellular layers: an outermost epithelium, middle stroma, and an innermost endothelium [2]. The cornea is highly innervated, and when healthy, it is avascular and immunologically privileged [1]. According to the World Health Organization, corneal blindness is the 4th largest cause of blindness worldwide [3]. In these cases, either loss of transparency or distortion of the refractive surface causes vision loss [1, 2]. Allograft cornea transplantation is the most common treatment.

The supply of high-quality donor cornea worldwide, however, falls well short of the demand. Allografting is contraindicated in a number of conditions such as autoimmune situations, chemical burns, and infections [4]. Allografting presents a very low but real risk of disease transmission, which is mostly circumvented by expensive screening procedures (e.g., processing fees in the USA are between $2,500 and $3,500 per cornea) [5]. These complications

Bernice Wright and Che J. Connon (eds.), *Corneal Regenerative Medicine: Methods and Protocols*, Methods in Molecular Biology, vol. 1014, DOI 10.1007/978-1-62703-432-6_2, © Springer Science+Business Media New York 2013

are compounded by the growing use of corrective eye surgery, which renders these corneas unsuitable for grafting, further reducing the availability of acceptable allogeneic supplies [6]. Initial success rates over 2 years are high, e.g., 85 % in a developed EU nation like Sweden [7]. The long-term success at 10–15 years, however, falls to 55 % [8], which is even lower than that for kidney transplantation [9]. In a developing country like India, even the short-term rates start out low, e.g., at 69 % in South India [10].

Replacement of allograft tissue has been a topic of interest, and there have been many versions of artificial replacements loosely grouped under "artificial corneas" that have been developed. These range from prostheses, known as keratoprostheses (KPro) to regenerative medicine approaches, to developing implants to promote regeneration of corneal tissues and nerves. Although decellularized corneas are becoming popular, we have not included them as they are not "artificially fabricated corneas."

This short chapter reviews a few examples taken from the spectrum of artificial corneas ranging from those in clinical use to those under clinical and laboratory investigation, and from prostheses to those designed to promote corneal regeneration.

2 Keratoprostheses in Clinical Use

The concept of an artificial cornea was first proposed by Guillaume Pellier de Quengsy in 1789 [11]. In 1856, Nussbaum implanted the first glass artificial cornea into a patient [12]. Since these early approaches, there have been numerous versions of KPro and today, some are at the research stages and several are under clinical investigation or in clinical use.

Since their first appearance in the 1800s, the core-and-skirt designs for artificial cornea still remain popular. The optical core now commonly comprises a transparent plastic such as poly (methyl methacrylate), while the porous material of the skirt varied. In the Boston KPro, one of the most successful KPros to date, the donor cornea is placed between the front and back plates, and the combination is sutured into the corneal opening of the patient. This construct is successful in restoring visual acuity and can persist in the eye of a patient for years [13]. Retroprosthetic membrane formation and persistent epithelial defects were the most common postoperative complications in 22 and 19 eyes out of 50 examined in a clinical study, respectively [14].

Another well-known KPro that is used clinically is the AlphaCor™. In this KPro, both optic and skirt were fabricated from poly (2-hydroxyethyl methacrylate) (PHEMA). The optic is solid but the skirts comprise interconnecting pores that allow biointegration with surrounding corneal tissue. In a recent study [15], the AlphaCor™ cornea was transplanted in 15 patients. The retention rates of the device after 1, 2, and 3 years were 87 %, 58 %, and 42 %, respectively. Stromal melt was the most significant complication for AlphaCor transplantation (nine cases). One attributable cause was that the PHEMA-based AlphaCor™ presents low permeability to glucose due to its low water content. Water content is a necessary property of the normal human cornea (80 %) to allow diffusion of nutrients to support a healthy corneal epithelium and hence, overall corneal health.

3 Keratoprostheses with Regenerative Capacities

To increase the water content of the artificial cornea, Jacob and coworkers incorporated methacrylic acid monomers into PHEMA together with cell adhesion peptides and various cytokines which permitted the achievement of water content of over 70 % and increased the permeability of the hydrogel [16]. KPro epithelization has been argued to be important for the reduction of postoperative complications (particularly infections) by restoring the natural cellular barrier of eyes [17]. Jacob and coworkers [16] also treated the surface of the artificial construct with laminin and fibronectin, and reported enhancement of corneal epithelial growth in vitro [16], as epithelial cells adhered to these proteins.

To aid bio-integration and encourage the adherence of epithelial cells, more recent work with KPro has included: the mixing of different hydrogel polymers to create copolymers that harness the beneficial properties of each component, tethering of biomolecules to the copolymers [16], and introduction of surface coatings comprising extracellular matrix motifs/molecules onto polyvinyl alcohol hydrogels [18].

One recent KPro was fabricated from a mechanically enhanced hydrogel material called "duoptix" to develop a new "core-and-skirt" model. It consisted of a double (interpenetrating) network of poly (ethylene glycol) and poly (acrylic acid) (PAG/PAA) in its central optic component that is surface modified for the growth of epithelial cells. Surrounding the optic is a microperforated rim designed to promote peripheral tissue integration with the host eye [19]. The team has also used a versatile photochemical surface modification strategy to site-specifically tether cell adhesion-promoting biomolecules to these otherwise non-adhesive hydrogels. A further innovation is the application of photolithographic patterning to the fabrication of the device which provides high level of control over the shape and structure of a hydrogel, and potential over the growth and differentiation of cells. A more recent innovation is the development of a single-piece KPro with their interpenetrating networked hydrogel, complete with photolithographed surface, which at the time of the report, was undergoing evaluation in animal models [20].

Another newer KPro, developed by the research group of Dr. Joachim Storsberg at the Fraunhofer Institute, Germany, comprises

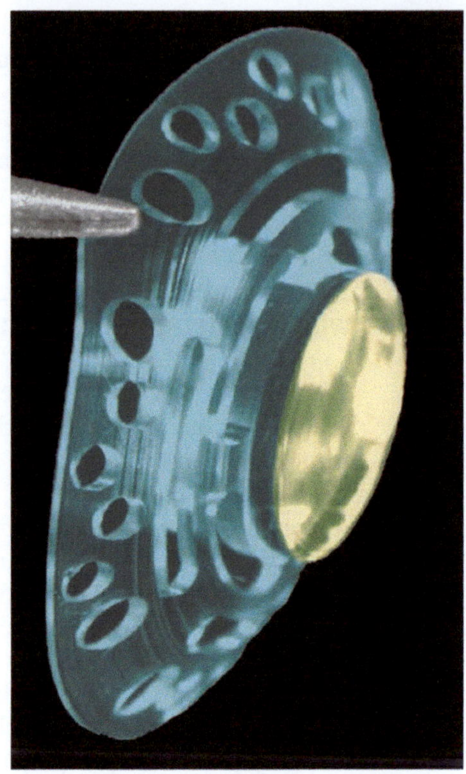

Fig. 1 The Fraunhofer epicorneal keratoprosthesis. This device comprises an anterior optic hydrophilic hydrogel (*yellow*), and an outer haptic cum cell-attracting skirt (*blue*) (photo from Storsberg et al. [21], with copyright permission from the author)

a cell-free transparent optic of polymethylmethacrylate and the fibronectin-coated haptic (Fig. 1). Results to date showed good epicorneal fixation and biointegration of device in animals; [21] and this device is now undergoing clinical testing in human subjects.

4 Collagen-Based Scaffolds and Templates for Corneal Regeneration

Hydrogels of collagen type I, the predominant biopolymer in the human cornea, are particularly attractive as matrix replacement scaffolds, partly because of their strength at relatively low concentrations, resulting from the virtually rigid rod properties of the collagen type I triple helix [22]. In addition, collagen contains the cell attachment motif arginine–glycine–glutamic acid [23]. Collagen hydrogels are, however, unstable and hence require stabilization, e.g., by chemical cross-linking [24] or plastic compression (*see* Chapters 9 and 10).

The authors and collaborators have tested a range of collagen-based biomaterials as corneal substitutes. Hybrid biosynthetic hydrogels based on collagen and a poly (*N*-isopropylacrylamide)-

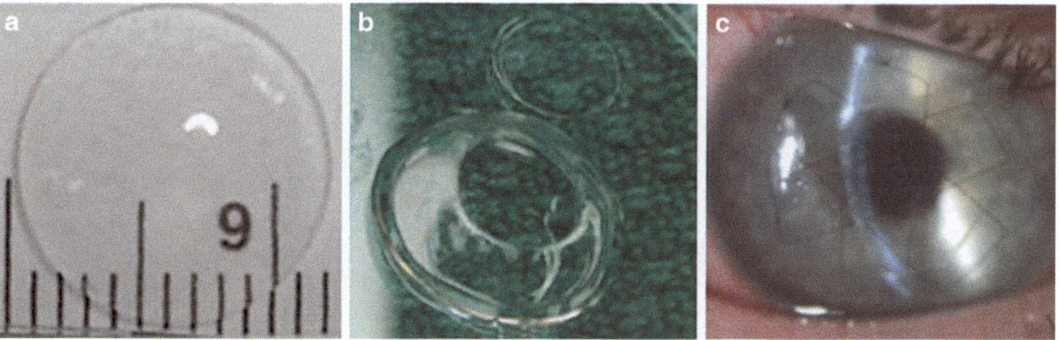

Fig. 2 (**a**) Cross-linked recombinant collagen-based implant. (**b**) The implant was trephined to prepare a button for corneal implantation. Damaged host tissue was removed to a similar depth and diameter and replaced by this button. (**c**) After implantation, the button was retained using overlying mattress sutures (from Fagerholm et al. [28], with the copyright permission from The American Association for the Advancement of Science)

derivative, acrylic acid, and *N*-acryloxysuccinimide grafted with tyrosine–isoleucin–glycine–serine–arginine (YIGSR) peptide were developed as mimic of the cornea stromal ECM-induced epithelial, stromal, and nerve regeneration when transplanted into the cornea of mini-pigs [25]. These implants had most likely emulated the ECM by allowing for cell–matrix interactions that resulted in the restoration of functional structures including the generation of a basement membrane between the implant and overlying epithelium, and stromal cell and nerve axon ingrowth. The incorporation of YIGSR into both bulk structure of the ethyl (dimethylaminopropyl) carbodiimide (EDC)/*N*-hydroxysuccinimide (NHS) cross-linked gels and onto the gel surface promoted the adhesion and proliferation of human corneal epithelial cells as well as neurite extension from dorsal root ganglia in vitro [26].

In 2009 and 2010, our group reported the 6-month and 2-year results, respectively, from implantation of EDC and NHS cross-linked recombinant human collagen corneal substitutes (*see* Chapter 10) in humans in a Phase I clinical study as lamellar grafts in ten patients [27, 28]. At 6 months postoperative, patients showed regeneration of epithelium, and ingrowth of stromal cells that was sufficient to anchor the implants [27]. After 2 years, the implants remained stably retained without the need for immunosuppression beyond the prophylactic use for the first 7 weeks after the surgery. Six of the ten patients had improved vision [28]. All ten showed corneal tissue, nerve and tear film regeneration, meaning that corneal epithelial cells grew over the implant, while stromal cell and nerves grew into the implant (Fig. 2), allowing for a tear film to form over the corneal surface. The tear film formation may have allowed the patients who were not able to tolerate contact lenses prior to the surgery to be able to now wear contact lenses to improve their eyesight.

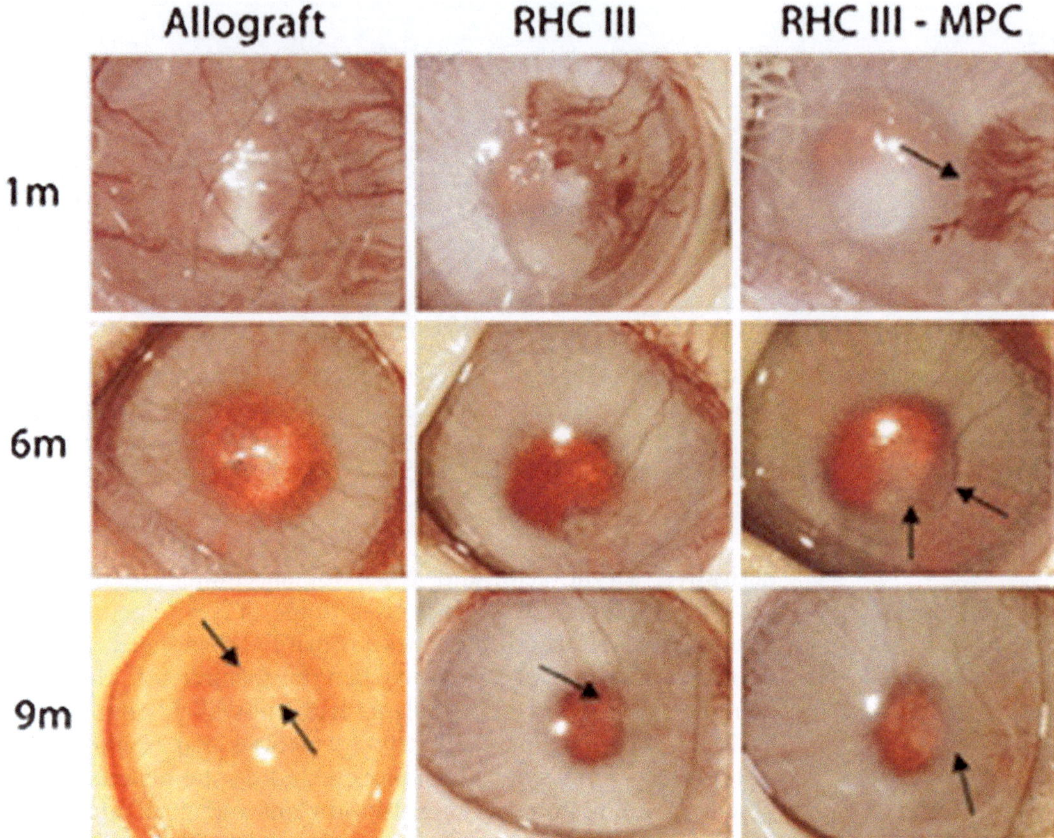

Fig. 3 Typical postoperative appearance of vascularized corneal implants. Serial images of the same corneas are presented. One month postoperatively, neovessels invaded implants, while RHCIII-MPC implants appeared to block vessels (*arrow*). By 6 months, vessels reduced in size and number, although central vessels remained in allograft and RHCIII implants while vessels in the RHCIII-MPC implant appeared to skirt the periphery before invading the implant (*arrows*). A further reduction in vascularization was noted at 9 months, but fine central vessels remained present (*arrows*) (photo from Hackett et al. [29], Copyright@ Association for Research in Vision and Ophthalmology)

More recent work from our laboratory has shown that the clinical implants, while successfully tested in patients with keratoconus and central scarring, were not sufficiently robust for implantation into rabbit models of severe corneal damage, e.g., alkali burn [29]. A second-generation RHC implant containing an integrated network of 2-methacryloyloxyethyl phosphorylcholine (MPC), a synthetic phosphorylcholine lipid (first used as antifouling coatings in arterial stents), was developed [30]. These optimized RHC-MPC implants are significantly tougher and more resistant to enzymatic degradation or neovascularization. When implanted into alkali-burnt corneas in rabbit models, the RHC-MPC implants were able to repel blood vessels and prevent neovascularization of the implant (Fig. 3), unlike the control allografts or RHC-only implants. These implants were shown to allow regeneration of the different corneal nerve subtypes when implanted as full-thickness grafts in guinea pigs [31].

5 Conclusions

There have been significant developments in tissue engineering approaches to replace the partial or full thickness of damaged or diseased corneas. Biomaterials have been developed to assist in these reparative procedures, to restore minimal function, or to regenerate the cornea to different degrees. For regeneration, biomaterials have been developed as interactive scaffolds to promote endogenous stem cell repair and regeneration. These different approaches may soon be able to supplement the supply of postmortem human corneas harvested for transplantation, or they may allow restoration of diseased or damaged corneas that cannot be treated by currently available techniques.

References

1. National Eye Institute. Facts about the cornea and corneal disease. www.nei.nih.gov/health/cornealdisease/#1
2. Asbell P, Brocks D (2010) Cornea overview. In Enclyclopedia of the Eye, DA Dartt, JC Besharse, R Dana. Academic Press, San Diego, USA, vol 1, pp. 552–531
3. WHO (2013) Prevention of blindness and visual impairment. Priority eye diseases. www.who.int/blindness/causes/priority/en/index9.html#
4. Hicks CR, Fitton JH, Chirila TV, Crawford GJ, Constable IJ (1997) Keratoprostheses: Advancing toward a true artificial cornea. Surv Ophthalmol 42:175–189
5. Hills K (2012) Challenges create cost increases for eye banks. www.revophth.com/content/d/cornea/c/32346
6. Mannis MJ, Sugar J (1995) Syphilis, serologic testing, and the setting of standards for eye banks. Am J Ophthalmol 119:93–95
7. Claesson M, Armitage WJ, Fagerholm P, Stevevi U (2002) Visual outcome in corneal grafts: a preliminary analysis of the Swedish Corneal Transplant Register. Br J Ophthalmol 86:174–180
8. Williams KA, Esterman AJ, Bartlett C, Holland H, Hornsby NB, Coster DJ (2006) How effective is penetrating corneal transplantation? Factors influencing long-term outcome in multivariate analysis. Transplantation 81:896–901
9. Wolfe RA (2004) Long-term renal allograft survival: a cup both half-full and half-empty. Am J Transplant 4:1215–1216
10. Dandona L, Naduvilth TJ, Janarthanan M, Ragu K, Rao GN (1997) Survival analysis and visual outcome in a large series of corneal transplants in India. Br J Ophthalmol 81: 726–731
11. Barnham JJ, Roper-Hall MJ (1983) Keratoprosthesis: a long-term review. Br J Ophthalmol 67:468–474
12. Chirila TV, Hicks CR, Dalton PD, Vijayasekaran S, Lou X, Hong et al. (1998) Artificial Cornea Prog Polym Sci 23:447–473
13. Aldave AJ, Kamal KM, Vo RC, Yu F (2009) The Boston type I keratoprosthesis: improving outcomes and expanding indications. Ophthalmology 116:640–651
14. Aldave AJ, Kamal KM, Vo RC, Yu F (2009) Epithelial debridement and Bowman's layer polishing for visually significant epithelial irregularity and recurrent corneal erosions. Cornea 28:1085–1090
15. Jiraskova N, Rozsival P, Burova M, Kalfertova M (2011) AlphaCor artificial cornea: clinical outcome. Eye 25:1138–1146
16. Jacob JT, Rochefort JR, Bi J, Gebhardt BM (2005) Corneal epithelial cell growth over tethered-protein/peptide surface-modified hydrogels. J Biomed Mater Res B Appl Biomater 72:198–205

17. Carlsson DJ, Li F, Shimmura S, Griffith M (2003) Bioengineered corneas: how close are we? Curr Opin Ophthalmol 14:192–197
18. Uchino Y, Shimmura S, Miyashita H, Taguchi T, Kobayashi H, Shimazaki J et al (2007) Amniotic membrane immobilized poly(vinyl alcohol) hybrid polymer as an artificial cornea scaffold that supports a stratified and differentiated corneal epithelium. J Biomed Mater Res B Appl Biomater 81:201–206
19. Myung D, Koh W, Bakri A, Zhang F, Marshall A, Ko J et al (2007) Design and fabrication of an artificial cornea based on a photolithographically patterned hydrogel construct. Biomed Microdevices 9:911–922
20. Myung D, Duchamel P-E, Cochran J, Noolandi J, Ta C, Frank C (2008) Development of hydrogel-based keratoprostheses: A materials perspective. Biotechnol Prog 24:735–741
21. Storsberg J, Kobuch K, Duncker G, Sel S (2011) Künstliche Augenhornhaut: biomaterialentwicklung eines ophthalmologischen Implantats mit biomimetischen Funktionalitaten. Dt Z Klin Forsch Ophthalmologie 5–6:58–61
22. Amis EJ, Carriere CJ, Ferry JD (1985) Effect of pH on collagen flexibility determined from dilute solution viscoelastic measurements. Int J Biol Macromol 7:130–134
23. Pierschbacher MD, Ruoslahti E (1987) Influence of stereochemistry of the sequence Arg-Gly-Asp-Xaa on binding specificity in cell adhesion. J Biol Chem 262:17294–17298
24. Hoffman AS (2002) Hydrogels for biomedical applications. Adv Drug Deliv Rev 54:3–12
25. Li F, Carlsson D, Lohmann C, Suuronen E, Vascotto S, Kobuch K et al (2003) Cellular and nerve regeneration within a biosynthetic extracellular matrix for corneal transplantation. Proc Natl Acad Sci U S A 100:15346–15351
26. Duan X, McLaughlin C, Griffith M, Sheardown H (2007) Biofunctionalization of collagen for improved biological response: scaffolds for corneal tissue engineering. Biomaterials 28:78–88
27. Fagerholm P, Lagali NS, Carlsson DJ, Merrett K, Griffith M (2009) Corneal regeneration following implantation of a biomimetic tissue-engineered substitute. Clin Transl Sci 2: 162–164
28. Fagerholm P, Lagali NS, Merrett K, Jackson WB, Munger R, Liu Y et al (2010) A biosynthetic alternative to human donor tissue for inducing corneal regeneration: 24-month follow-up of a phase 1 clinical study. Sci Transl Med 2:46–61
29. Hackett JM, Lagali N, Merrett K, Edelhauser H, Sun Y, Gan L et al (2011) Biosynthetic corneal implants for replacement of pathologic corneal tissue: performance in a controlled rabbit alkali burn model. Invest Ophthalmol Vis Sci 52:651–657
30. Liu W, Deng C, McLaughlin CR, Fagerholm P, Lagali NS, Heyne B et al (2009) Collagen-phosphorylcholine interpenetrating network hydrogels as corneal substitutes. Biomaterials 30:1551–1559
31. McLaughlin CR, Acosta MC, Luna C, Liu W, Belmonte C, Griffith M et al (2010) Regeneration of functional nerves within full thickness collagen-phosphorylcholine corneal substitute implants in guinea pigs. Biomaterials 31:2770–2778

Part II

Corneal Stem Cell Enrichment and Characterization

Chapter 3

Clonal Analysis of Limbal Epithelial Stem Cell Populations

Ursula Schlötzer-Schrehardt

Abstract

While convincing data clearly suggest the presence of stem cells in the basal limbal epithelium in vivo, testing the proliferation, self-renewal, and differentiation capacity of stem cells relies on the development of methodologies that allow for their isolation and extensive propagation in vitro. Clonal analysis involving differentiation between short-lived transient cell clones and long-lived stem cell clones is an invaluable technique to identify stem cells in vitro, and allows cells to be expanded over multiple passages. This chapter describes a protocol for the isolation, expansion, and clonal analysis of limbal epithelial stem cells. The cultivation method described may be essential for long-term restoration of the damaged ocular surface in patients with limbal stem cell deficiency.

Key words Stem cells, Transient amplifying cells, Colony forming efficiency, Clonal analysis, Limbal epithelial stem cells, Holoclones, Meroclones, Paraclones, Feeder cells

1 Introduction

Cultures of limbal epithelial stem cells have long been used to prepare grafts that can restore epithelial defects of the human cornea [1]. An essential prerequisite of epithelial grafts for long-term restoration of the ocular surface is the presence of an adequate number of stem cells [2]. Stem cells are defined by their ability to self-renew and differentiate into various cell types. In contrast, lineage-restricted progenitors, i.e., transient amplifying cells, exhibit limited potential for proliferation/self-renewal. Distinguishing true stem cells from progenitors is a fundamental goal in stem cell biology and has significant implications for therapeutic applications.

Since the pioneering work by Rheinwald and Green [3], studies have shown that long-term survival and serial expansion of epithelial stem cells are possible if these cells are cocultured with embryonic fibroblast feeder cells. Under these conditions, which may compensate for the lack of a niche microenvironment, cells with clonogenic ability can give rise to a clonal colony representing the progeny of a single cell. Clonal analysis of human epidermal keratinocytes cultured on feeder layers has identified three

Bernice Wright and Che J. Connon (eds.), *Corneal Regenerative Medicine: Methods and Protocols*, Methods in Molecular Biology, vol. 1014, DOI 10.1007/978-1-62703-432-6_3, © Springer Science+Business Media New York 2013

types of clonogenic cells, holoclones, meroclones, and paraclones, with different capacities for proliferation [4]. Holoclone-forming cells have all the hallmarks of stem cells, including self-renewing capacity and a great proliferative potential, whilst meroclones and paraclones are generated by different stages of transient amplifying progenitors with a limited capacity for proliferation. Thus, only holoclones possess regenerative properties and the potential for self-renewal. Holoclones represent approximately 1–5 % of clonogenic keratinocytes, the vast majority of which generate meroclones and paraclones.

The discovery of clonogenic keratocytes was followed by the identification of holoclone-forming cells in the limbal epithelium and the development of a culture system that involves enrichment of limbal stem cells (LSC) by clonal growth on a feeder layer before they are seeded onto transplantable carriers to produce epithelial sheets [5]. Consistently, keratinocytes cultured using this method have been used to permanently restore the corneal surface of patients with total limbal stem cell deficiency (LSCD) [6], and demonstration of p63-positive holoclones within epithelial grafts has been considered as a means of quality control for transplantation [7, 8]. More recently, a simplified culture system supporting the enrichment and preservation of limbal stem and progenitor cells by using clonal expansion was proposed [9].

Therefore, clonal analysis provides a means to sort heterogeneous stem and progenitor cell populations ex vivo and define cells with true stem cell properties. Provision of suitable cellular microenvironments, such as embryonic fibroblasts, increases stem cell survival after removal from their niche and dispersal into single cells in vitro. Using those approaches, it has been possible to determine the frequency of stem cells capable of clonogenic growth as a functional measure of stem cell self-renewal at the single cell level [10]. Here, we describe the performance of clonal assays by scoring colonies according to their proliferative and self-renewing capacity and according to morphological criteria. These methodological approaches, which require the presence of a feeder layer, proper serum concentrations, defined calcium concentrations, and addition of growth factors, can be used to generate corneal epithelial cell sheets which support the survival of LSC in the cultured graft.

2 Materials

Perform all procedures in a class II biological safety cabinet unless indicated otherwise.

2.1 Cell Lines

Swiss albino mouse 3T3 fibroblasts (ACC 173; DSMZ, Braunschweig, Germany, or ATCC-CRL-1658; LGC Standards, Wesel, Germany).

2.2 Cell Culture

1. Culture apparatus: Flasks, 6-well plates, LabTek Chamber Slides.
2. Cloning cylinders, glass: 8 or 10 mm in diameter, 976 V silicone high vacuum grease.
3. Human corneal growth supplement (HCGS) containing 0.18 μg/mL hydrocortisone, 5 μg/mL insulin, 5 μg/mL transferrin, 1 ng/mL epidermal growth factor (EGF), and 0.2 % (w/v) bovine pituitary extract.
4. Culture medium I: Dulbecco's modified Eagle's medium (DMEM) containing 4,500 mg/L glucose, L-glutamine, and sodium pyruvate, supplemented with 10 % (w/v) fetal calf serum (FCS), and 1.5 % (v/v) penicillin/streptomycin/amphotericin B solution.
5. Culture medium II: MCDB 151 medium supplemented with HCGS, 5 ng/mL human EGF, 10 % (w/v) FCS, and 5 μg/mL gentamycin. MCDB 151 medium contains low calcium concentrations (0.03 mM).
6. Culture medium III: Three parts DMEM without calcium and one part Ham's F12 supplemented with 500 mg/L L-glutamine, HCGS, 5 ng/mL hEGF, 10 % (w/v) FCS, and 5 μg/mL gentamycin. A final calcium concentration of 0.4 mM is achieved by using a 0.2 M $CaCl_2$ solution.
7. Dispase II (neutral protease, grade II; 1 U/mg): Dissolve 125 mg in 50 mL culture medium (2.5 U/mL); aliquot and freeze into one time-use volumes (e.g., 2 mL).
8. Trypsin/ethylenediaminetetraacetic acid (EDTA): 0.25 % (w/v) trypsin–0.02 % (w/v) EDTA.
9. Trypsin/EDTA: 0.05 % trypsin–0.02 % EDTA.
10. Versene: 1:5,000 (1×).
11. Gentamycin.
12. Penicillin–streptomycin–amphotericin B (100×).
13. Mitomycin C (2 mg): Dissolve in 2 mL H_2O, aliquot and freeze stock solution (1 mg/mL) into one time-use volumes (e.g., 50 μL), thaw immediately prior to use, and prepare working dilution by adding 50 μL stock solution to 10 mL medium.
14. Rhodamine B: 2 % (w/v) in H_2O.
15. Phosphate-buffered saline (PBS) without Ca^{2+} and Mg^{2+}.
16. PBS with Ca^{2+} and Mg^{2+}.
17. Paraformaldehyde: 4 % (w/v) in PBS.

2.3 Antibodies

1. Rabbit anti-human Bmi-1 (Abcam): 1:500 dilution in PBS.
2. Rabbit anti-human p63-α (Cell Signalling): 1:100 dilution in PBS.
3. Alexa 488- or Alexa 555-conjugated goat anti-rabbit IgG (H+L) (Invitrogen): 1:300 dilution in PBS.

2.4 Equipment

1. Phase contrast microscope.
2. Image analysis system (Cell^F: Olympus Soft Imaging Solutions GmbH, Münster, Germany).
3. Cell counting system (CASY: Roche, Basel, Switzerland).
4. Desk scanner.

3 Methods

3.1 Preparation of Feeder Layer

1. Grow 3T3 fibroblasts in 250 mL culture flasks in culture medium I at 37 °C under 5 % CO_2 and 95 % humidity.
2. Always maintain cells at the stage of subconfluence and passage them using 0.05 % (w/v) trypsin–0.02 % (w/v) EDTA at a ratio of 1:5–1:10 when approximately 80 % confluence is reached (*see* **Note 1**).
3. For mitotic inactivation, treat cells at 70–80 % confluence with 5 μg/mL mitomycin C in culture medium I for 2 h at 37 °C.
4. Remove and properly dispose of the inactivation medium (*see* **Note 2**).
5. Wash cells three times with PBS (without Ca^{2+} and Mg^{2+}) and dissociate them with 0.05 % (w/v) trypsin–0.02 % (w/v) EDTA for 2 min at 37 °C.
6. Add equal amounts of culture medium I, transfer cells to a 15 mL centrifuge tube, and centrifuge for 10 min at 179.998 × *g*.
7. Count cells and plate them into 6-well plastic dishes or 2-well chamber slides at a density of 2×10^4 cells/cm².
8. Incubate overnight and use as feeder layers the next day and up to day 6. Replenish medium every other day.

3.2 Limbal Epithelial Cell Culture

1. Take small biopsies (2 × 2 mm²) from the limbus of normal donor corneas preferentially from the superior region.
2. Incubate limbal biopsies in dispase II (2.5 U/mL) in culture medium II or III for 1.5 h at 37 °C (*see* **Note 3**).
3. Gently scrape off the epithelial cell sheet with a spatula or the back side of curved forceps and incubate in 0.25 % (w/v) trypsin–0.02 % (w/v) EDTA for 10 min at 37 °C to obtain single-cell suspensions.
4. Rapidly block enzymatic activity with a double volume of culture medium II or III containing 10 % (w/v) FCS to prevent cell damage.
5. Transfer cells to a 15 mL centrifuge tube and centrifuge for 10 min at 179.998 × *g*.

6. Remove supernatant and gently resuspend cells in growth medium II or III.
7. A single-cell suspension is obtained by gentle pipetting to help dislodge the cells.
8. Seed single-cell suspension at a density of 1×10^3 viable cells/cm^2 on a 1-day-old 3T3 feeder cell layer in 6-well culture plates.
9. Incubate cell cultures at 37 °C under 5 % CO_2 and 95 % humidity in culture media II or III (*see* **Note 4**).
10. After a 24-h incubation period, gently remove unattached cells by aspiration and maintain attached cells in culture for approximately 12–14 days.
11. Change medium every 2–3 days.
12. After 1 week in culture, monitor colony formation daily using an inverted phase contrast microscope (*see* **Note 5**).

3.3 Clonal Analysis (See Note 6)

1. Seed single cells at a density of 1×10^3 viable cells/cm^2 on a 1-day-old 3T3 feeder cell layer in 6-well culture plates.
2. After 10–12 days of culture, identify clones using an inverted phase contrast microscope (*see* **Notes** 7 and **8**).
3. Mark individual large colonies (presumed holoclones), which are well isolated from other colonies, by drawing a circle around them on the bottom of the dish with a marking pen.
4. Remove the growth medium and harvest individual colonies by using cloning cylinders which have been dipped into sterile silicone grease and gently placed over a colony.
5. Add about 0.2 mL of 0.25 % (w/v) trypsin–0.02 % (w/v) EDTA to the cloning cylinder and incubate for approximately 5–10 min at 37 °C.
6. Rapidly block enzymatic activity by adding a few drops of culture medium containing 10 % (w/v) FCS, transfer cells to a 15 mL centrifuge tube or a 1.5 mL microtube, and centrifuge for 10 min at 179.998 × *g*.
7. Count cells and transfer them at a density of 1×10^3 cells/cm^2 into three wells of a new 6-well plate with feeder layer (*see* **Note 9**). Incubate cell cultures at 37 °C under 5 % CO_2 and 95 % humidity in culture media II or III.
8. After approximately 12 days from plating, with cells at 70–80 % confluence, use two of the three wells for serial propagation and determination of population doublings: The population doubling value (X) of the clones is obtained using the formula $X = 3.322 \log (N/N_0)$, where N is the total number of cells obtained at each passage and N_0 is the number of clonogenic, i.e., colony-forming, cells. Cloning and subcloning are

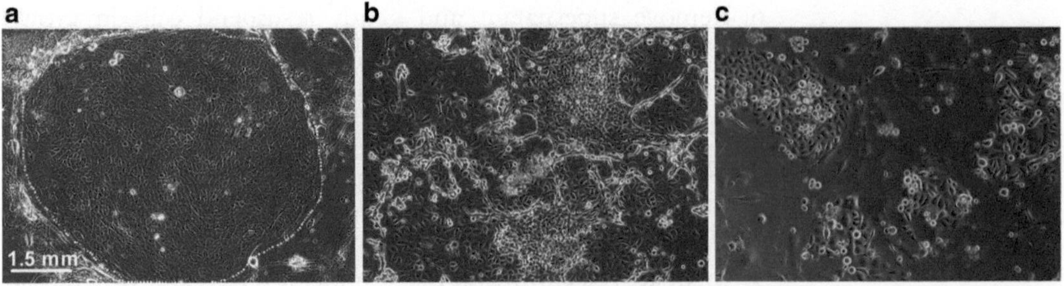

Fig. 1 Characterization of clonal stages of limbal epithelial cells. Phase contrast images of holoclones, meroclones, and paraclones generated by LSC cultivated using a 3T3 feeder layer

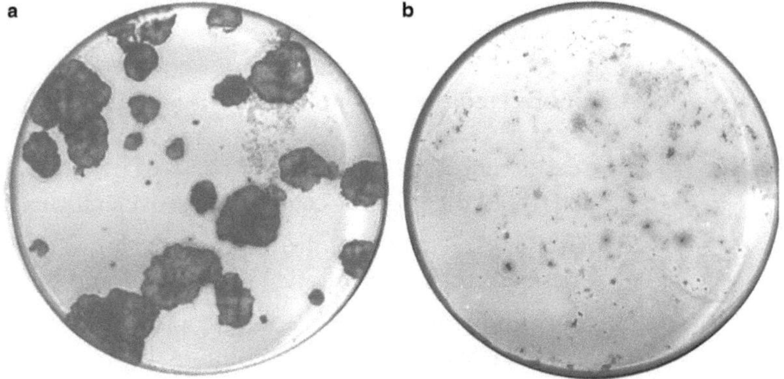

Fig. 2 Colony forming efficiency of limbal epithelial stem cells (LESC). LSC populations containing holoclones display large colonies (**a**) and those mainly consisting of paraclones display aborted colonies (**b**)

performed by passaging cells at 70–80 % subconfluence at a strict time schedule.

9. Use one of the three wells (indicator well) for classification of clonal type (Fig. 1).
10. Remove the feeder layer with Versene or 0.02 % (w/v) EDTA (2 mL/well) for 30 s.
11. Rinse with PBS (without Ca^{2+} and Mg^{2+}) under vigorous pipetting and microscopic control to completely detach the feeder cells.
12. Fix colonies in 4 % (w/v) paraformaldehyde in PBS for 2 h and rinse with PBS.
13. Stain colonies with 2 % (w/v) rhodamine B in H_2O (2 mL/well) for 15 min and rinse with water (Fig. 2).
14. Scan plate with a desk scanner or score colonies under a dissecting microscope and classify them into holo-, mero-, and paraclones according to their size and morphology (*see* **Note 8** and Fig. 2).

15. Determine colony size, defined as diameter of individual clones in millimeter, and colony density, defined as percentage area of the culture dish covered by all holoclones, by using an image analysis software (e.g., Cell^F). The number of aborted colonies (paraclones) is used to define the original clone as holoclone, meroclone, or paraclone (*see* **Note 6**).
16. Calculate the colony forming efficiency (CFE) as the number of all colonies (holo-, mero-, and paraclones) divided by the total number of cells seeded per well × 100 % (*see* **Note 10**).

3.4 Phenotypic Characterization of Colonies (See Note 11)

1. For immunocytochemistry, seed single-cell suspension at a density of 1×10^3 cells/cm^2 on a 3T3 feeder cell layer in glass chamber slides.
2. Culture cells in medium II or III for approximately 12 days.
3. Monitor colony formation by inverted phase contrast microscopy.
4. Remove growth medium and fix colonies with cold acetone for 10 min at 4 °C.
5. After washing with PBS (with Ca^{2+} and Mg^{2+}), permeabilize cells with 0.1 % (w/v) Triton X-100 in PBS for 10 min at 4 °C.
6. Block unspecific binding with 10 % (v/v) normal serum for 30 min at room temperature.
7. Incubate cells in primary antibodies against p63α or Bmi-1 diluted in PBS for 2 h at room temperature or overnight at 4 °C, respectively.
8. Rinse with PBS and visualize antibody binding by Alexa 488- or Alexa 555-conjugated secondary antibodies diluted in PBS for 30 min at room temperature.
9. Rinse with PBS and perform nuclear counterstaining with propidium iodide or DAPI for 10 min at room temperature.
10. After washing, pipette mounting medium over tissue sections on slides and cover with a glass coverslip.
11. Evaluate slides with a fluorescence microscope and count the number of p63α- or Bmi1-positive nuclei.
12. Include appropriate negative control experiments, i.e., replace the primary antibody by PBS or equimolar concentrations of an isotype control antibody.

3.5 Subculture of Clonal Cells

1. After approximately 21 days of culture, remove the feeder layer with 0.02 % (w/v) EDTA or Versene for 30 s.
2. Harvest holoclones by trypsinization (0.25 % (w/v) trypsin–EDTA) for 15 min at 37 °C.
3. Seed cells on top of a carrier, i.e., amniotic membrane, fibrin gel, and collagen membrane, at a density of 1×10^5 cells/cm^2.

4. Cultivate cells in medium II or medium III for 14–16 days (*see* **Note 12**).
5. Dissociate cells from the substrate.
6. Calculate the CFE to determine the percentage of stem cells persisting after subculture.

4 Notes

1. Mouse 3T3 cells require careful culture and maintenance. It is important to keep the cells in a healthy proliferating state to ensure production of factors needed for LSC growth. They therefore need to be carefully monitored to avoid confluency and overgrowth of cell cultures, which results in early senescence.
2. Mitomycin C inhibits DNA synthesis and cell division. It is toxic and possibly carcinogenic. Therefore handle with caution by wearing protective gloves and safety glasses, and work in an area with good ventilation. Read the material safety datasheet from the provider for proper disposal methods.
3. Dispase is not inhibited by serum.
4. The concentration of calcium found in classic culture media is usually rather high (1–2 mM). As calcium affects cell signalling and cell differentiation, calcium levels are generally lower in media developed for clonal cell culture. MCDB 151, developed for keratinocytes, contains very low levels of calcium (0.03 mM), whereas DMEM/F12 was adjusted to medium levels of calcium (0.4 mM).
5. Although the CFE is usually similar in both types of culture media, the rate of colony growth will vary. In MCDB 151 medium, macroscopic colonies will appear within 10–12 days after inoculation under a phase contrast microscope, and after 3–4 weeks, the colonies eventually fuse and generate a stratified layer. Colony growth is slightly faster in DMEM/F12 medium with macroscopic colonies already appearing after 7–9 days.
6. This method of clonal analysis, which has been originally described by Barrandon and Green [4] and further elaborated by Pellegrini et al. [5], scores individual clones according to their size, cell number, morphology, and, most importantly, the percentage of abortive colonies of their progeny. By using this method, each individual single cell-derived colony is transferred to three dishes: two dishes are used for serial propagation and further analysis; the third (indicator) dish is fixed 12 days later and stained with rhodamine B for classification of the clonal type (holoclone, meroclone, or paraclone) and the CFE.

   The clonal type is determined by the percentage of aborted colonies formed in the indicator dish: When 0–5 % of colonies are abortive clones, the original clone is scored as holoclone. When all colonies formed are abortive clones (or when no colonies formed), the original clone is classified as paraclone. When >5 % and <100 % of the colonies are terminal, the original clone is classified as meroclone. In addition, one of the other two dishes is used for the calculation of population doublings generated during serial cultivation.

7. Non-stem cells generate clones known as transient clones which tend to be variable in size, are short-lived, and are mature to contain only differentiated cells. In contrast, stem cell clones tend to be larger and more uniform in size, persist longer, and always contain undifferentiated, intermediate, and differentiated progeny. Distinguishing short-lived transient clones from long-lived stem cell clones is the key step in identifying stem cells. Hence the time elapsed between clone initiation and analysis is the key variable.

   At short intervals, there are likely to still be many transient clones, while long intervals support the evolution of stem cell clones. Very long time points may be less useful, because many stem cell clones will likely have turned over. At short periods after clone induction, transient clones will predominate; they may be irregular in shape due to cell migration. Eventually, after the transient clones have turned over, mostly stem cell clones will be present. They are likely to be uniform in size.

8. Three types of clones can be observed: holoclones (large colonies, in which the cells are capable of prolonged self-maintenance, when subcloned); meroclones (only a limited proportion of cells from a clone can be serially subcloned, and the emerging colonies are of differing size reflecting highly variable proliferative potentials); and paraclones (form no colonies or very few small colonies which rapidly abort after the first subcloning).

   Although it is not possible to distinguish holoclones with certainty from meroclones solely based on morphologic parameters in short-term clonogenic assays, morphological criteria can be used for a preliminary estimate: large, nearly circular colonies (4–10 mm in diameter) with smooth outlines are usually identified as holoclones; medium-sized colonies (1–4 mm in diameter) with wrinkled outlines as meroclones; and small highly irregular colonies (<1 mm in diameter) as paraclones (Fig. 1). Holoclones consist of small cuboid, densely packed cells in a regular mosaic pattern, concentrated in the periphery of the colony. In contrast, most cells in the meroclones and all cells in the paraclones have an increased cell size and appear irregular and flattened.

9. Holoclones contain $2–5 \times 10^4$ cells/mm^2 on average.

10. Average CFE ranges from 0.3 to 0.4 %.

11. Immunodetection of the transcription factor p63, particularly its isoform ΔNp63α, and Bmi-1 has been described as an important method for identification of stem cell-derived holoclones and the presence of stem cells within a cultured graft [2, 8].
12. For medium II, switch culture conditions from low-calcium conditions to medium-calcium conditions (0.4 mmol/L calcium) by adding $CaCl_2$ to the medium to promote cell adhesion.

References

1. Pellegrini G, Traverso CE, Franzi AT, Zingirian M, Cancedda R, De Luca M (1997) Long-term restoration of damaged corneal surfaces with autologous cultivated corneal epithelium. Lancet 349:990–993
2. Pellegrini G, Rama P, De Luca M (2011) Vision from the right stem. Trends Mol Med 17:1–7
3. Rheinwald JG, Green H (1975) Serial cultivation of strains of human epidermal keratinocytes: the formation of keratinizing colonies from single cells. Cell 6:331–343
4. Barrandon Y, Green H (1987) Three clonal types of keratinocyte with different capacities for multiplication. Proc Natl Acad Sci U S A 84: 2302–2306
5. Pellegrini G, Golisano O, Paterna P, Lambiase A, Bonini S, Rama P et al (1999) Location and clonal analysis of stem cells and their differentiated progeny in the human ocular surface. J Cell Biol 145:769–782
6. De Luca M, Pellegrini G, Green H (2006) Regeneration of squamous epithelia from stem cells of cultured grafts. Regen Med 1:45–57
7. Pellegrini G, Dellambra E, Golisano O, Martinelli E, Fantozzi I, Bondanza S et al (2001) p63 identifies keratinocyte stem cells. Proc Natl Acad Sci U S A 98:3156–3161
8. Di Iorio E, Barbaro V, Ruzza A, Ponzin D, Pellegrini G, De Luca M (2005) Isoforms of DeltaNp63 and the migration of ocular limbal cells in human corneal regeneration. Proc Natl Acad Sci U S A 102:9523–9528
9. Meyer-Blazejewska EA, Kruse FE, Bitterer K, Meyer C, Hofmann-Rummelt C, Wünsch PH et al (2010) Preservation of the limbal stem cell phenotype by appropriate culture techniques. Invest Ophthalmol Vis Sci 51: 765–774
10. Hope K, Bhatia M (2011) Clonal interrogation of stem cells. Nat Methods 8:S36–S40

Chapter 4

The Use of Bromodeoxyuridine Incorporation Assays to Assess Corneal Stem Cell Proliferation

Ashley M. Crane and Sanjoy K. Bhattacharya

Abstract

Bromodeoxyuridine (BrdU) incorporation assays have long been used to detect DNA synthesis in vivo and in vitro. The key principle of this method is that BrdU incorporated as a thymidine analog into nuclear DNA represents a label that can be tracked using antibody probes. In this chapter, we describe BrdU incorporation into limbal stem cells. The colorimetric reaction produced by this assay can be detected by immunohistochemistry, and using appropriate controls, it can be used for determination of proliferating properties of restricted progenitor cells derived from the cornea.

Key words Bromodeoxyuridine, 5-Bromo-2′-deoxyuridine, Thymidine analog, Bromodeoxyuridine incorporation assay

1 Introduction

5-Bromo-2′-deoxyuridine (bromodeoxyuridine (BrdU)) is a thymidine analog which differs from thymidine in its substitution of bromine for a methyl group. BrdU has been in use since the 1980s to label proliferating and migrating cells in various human and animal organs [1]. Since this time, BrdU incorporation has been reported as a method to detect DNA synthesis in over 20,000 scientific studies [2]. The use of this halopyrimidine is preferred over the older technique involving the use of ^{3}H-Thymidine (^{3}H-dT) autoradiography, due to the lower cost, faster processing time, and nonradioactive properties of BrdU [3]. Another advantage of using BrdU for labeling is that although this chemical can be considered to be toxic [3], it is usually not so at the low concentrations used for labeling [4].

BrdU competes with thymidine for incorporation into nuclear DNA during the S-phase of the cell cycle [1]. It is important, however, to note that BrdU can also be incorporated into nuclear DNA during cell repair or cell degeneration [5], but the amount of analog

Bernice Wright and Che J. Connon (eds.), *Corneal Regenerative Medicine: Methods and Protocols*, Methods in Molecular Biology, vol. 1014, DOI 10.1007/978-1-62703-432-6_4, © Springer Science+Business Media New York 2013

expected to be incorporated into DNA in these situations is much lower than that expected during cell division [5]. Therefore, BrdU serves as a marker of DNA synthesis, and separate means such as counting mitotic figures may be employed to ensure accuracy with regard to cell division [5, 6]. After incorporation into nuclear DNA, BrdU can be detected with immunohistochemical methods. One limitation of this method is that, unlike ^{3}H-dT autoradiography, BrdU immunohistochemistry is not stoichiometric. That is, BrdU labeling relies heavily on the accuracy of detection techniques and therefore cannot be used for absolute quantification [5].

After BrdU is incorporated into nuclear DNA, samples are fixed and incubated with anti-BrdU monoclonal antibodies and nucleases (or they are exposed to heat or other conditions which cause DNA denaturation). This denaturation is necessary to allow anti-BrdU monoclonal antibodies to gain access to the incorporated BrdU in single stranded DNA. The sample is subsequently incubated with a secondary antibody against the anti-BrdU antibody. The secondary antibody is often associated with an enzyme, which when exposed to its added substrate, causes a colorimetric reaction. The reaction can be observed by bright-field microscopy.

BrdU has a history of use in corneal limbal stem cells, and has been applied in vivo in both dividing and slow cycling limbal cells [7]. The thymidine analog has also been used in double labeling protocols in limbal cells [8] and therefore can be combined with other experiments designed to detect these cells. BrdU incorporation assays can be used to investigate the proliferation of adherent and suspension cells as well as cell cycling in tissue sections.

2 Materials

Store all materials at room temperature or as recommended by the manufacturer. Dispose of materials using established disposal protocols.

1. Ethanol fixative: add 50 mM glycine solution to 70 mL absolute ethanol for a total of 100 mL of ethanol fixative, adjust to pH 2.0, and store at 4 °C.
2. Substrate buffer: prepare 100 mM Tris–HCl-buffer, 100 mM NaCl, 50 mM $MgCl_2$, adjust to pH 9.5 (at 20 °C), and store at 4 °C.
3. Sterile culture medium (media that is typically used to culture limbal stem cells are discussed in Chapters 3 and 6).
4. Plastic slide box.
5. Phosphate buffered saline (PBS) (1×).
6. Double distilled water.

7. Glycerol gelatin aqueous slide mounting medium (cat. no. GG1, Sigma Aldrich).
8. Fat-free, poly-L-lysine coated glass slides for frozen tissue sections.
9. Cryostat for preparing tissue sections.
10. BrdU Labeling and Detection Kit II (cat. no. 11299964001, Roche Diagnostics) containing: (a) BrdU labeling reagent (1,000× concentrate). For in vitro labeling, prepare BrdU labeling reagent immediately prior to use. Dilute BrdU labeling reagent with sterile cell culture medium at 1:1,000 to create BrdU labeling solution. Do not store. For in vivo labeling, do not dilute BrdU labeling reagent. Prepare 2 mL/100 g body weight of the animal. (b) Washing buffer concentrate (10× concentrate). Dilute washing buffer concentrate with double distilled water at 1:10 to create working washing buffer. Store at 4 °C. (c) Incubation Buffer. (d) Anti-BrdU antibody containing nucleases for DNA denaturation. Prepare anti-BrdU antibody immediately prior to use, and dilute with incubation buffer at 1:10 to create a working solution. Do not store. (e) Anti-mouse Ig-alkaline phosphatase (AP). Prepare anti-mouse Ig-AP immediately prior to use, and dilute with PBS at 1:10 to create a working solution. Do not store. (f) NBT/BCIP. Prepare NBT/BCIP immediately prior to use. Combine 13 μL NBT and 10 μL BCIP solution with 3 mL substrate buffer. This will generate a colored substrate solution. Do not store.

3 Methods

3.1 Immunocytochemistry for Adherent Cells

1. Grow limbal stem cells on slides until 50 % confluent (*see* Chapters 3, 7, and 9).
2. Add BrdU labeling solution to cover the slides and incubate the cells at 37 °C, 5% CO_2. To detect cell proliferation, incubate the cells for 60 min. To detect cell repair, incubate the cells for 30 min (*see* **Note 1**).
3. Wash slides three times with diluted washing buffer.
4. Add anti-BrdU antibody working solution to cover the slides and incubate for 30 min at 37 °C (*see* **Notes 2** and **3**).
5. Wash slides three times with diluted washing buffer.
6. Add anti-mouse Ig-AP working solution to cover slides and incubate for 30 min at 37 °C.
7. Wash slides three times with diluted washing buffer.
8. Add NBT/BCIP substrate solution to cover the slides and incubate for 30 min at 25 °C.
9. Mount slides and inspect with a bright-field microscope.

3.2 Immunohisto-chemistry for In Vivo Frozen Tissue Sections

Appropriate ethical approval must be obtained before undertaking research using animals as experimental models.

1. Introduce the stock BrdU labeling reagent (2 mL reagent/100 g animal body weight) intravenously into the animal.
2. Sacrifice the animal after administering the BrdU labeling reagent.
3. Prepare 5 μm thick sections of limbus with a cryostat and place on poly-L-lysine glass slides.
4. Fix limbal tissue in ethanol.
5. Wash frozen sections three times with diluted washing buffer and air-dry the area to be stained.
6. Add anti-BrdU antibody working solution to cover the slides and incubate for 30 min at 37 °C.
7. Wash slides three times with diluted washing buffer and air-dry the area to be stained.
8. Add anti-mouse Ig-AP working solution to cover the slides and incubate for 30 min at 37 °C.
9. Wash slides three times with diluted washing buffer and air-dry.
10. Add NBT/BCIP substrate solution to cover the slides and incubate for 30 min at 25 °C.
11. Wash slides three times with diluted washing buffer and air-dry.
12. Mount slides and inspect with a bright-field microscope.

3.3 Immunohisto chemistry for Tissue Slices Prepared by Frozen Section

1. Cut tissue samples to produce slices that are approximately 1 mm thick.
2. Add BrdU labeling solution to cover the cut sample and incubate the cells at 37 °C, 5% CO_2. To detect cell proliferation, incubate the cells for 60 min. To detect cell repair, incubate the cells for 30 min.
3. Place the sample in washing buffer and incubate for 30 min at 37 °C.
4. Repeat **steps 3–12** of Subheading 3.2.

4 Notes

1. As previously mentioned, the use of BrdU does come with some limitations. The use of BrdU in vivo has been shown to adversely affect cell survival, migration, and final position in the brain [3], and may similarly affect limbal stem cells to an unknown extent. Furthermore, with some labeling kits such as

BrdU Labeling and Detection Kit II (Roche Diagnostics) there is 10 % cross-reactivity with 5-iodo-2′deoxy-uridine.

2. The anti-BrdU monoclonal antibody requires DNA denaturation because it is directed against single stranded DNA [1]. Methods used for denaturation, including heat and nucleases, threaten DNA integrity. Recent new methods of detecting DNA synthesis in dividing cells include the use of another thymidine analog, 5-ethynyl-2′-deoxyuridine (EdU) and the use of fluorescent azides and "click" chemistry in in vitro models [9, 10]. This method has the advantage of not requiring DNA denaturation and therefore preserves DNA structure [9, 10]; however, this method has been observed to be highly cytotoxic in some cell lines [11]. Similarly, the deoxycytidine analog 5-Ethynyl-2′-deoxycytidine (EdC), does not require DNA denaturation. This method is thought to be less cytotoxic than EdU and has been used in vitro and in vivo [12].
3. The fact that BrdU should be used in conjunction with other methods to detect DNA proliferation also hampers its utility. When this proves to be limiting, the use of techniques such as Ki67 monoclonal antibodies can be considered. This antibody detects all phases of proliferating cells, but not cells undergoing repair [13], and can therefore avoid the need for additional experiments or counting mitotic figures.

Acknowledgments

This work was supported by DOD grant W81XWH-09-1-0674 (Project 2.2), a career award and unrestricted funds from Research to Prevent Blindness and NIH core grant P30-EY14801.

References

1. Miller MW, Nowakowski RS (1988) Use of bromodeoxyuridine-immunohistochemistry to examine the proliferation, migration and time of origin of cells in the central nervous system. Brain Res 457:44–52
2. Cavanagh BL, Walker T, Norazit A, Meedeniya AC (2011) Thymidine analogues for tracking DNA synthesis. Molecules 16:7980–7993
3. Duque A, Rakic P (2011) Different effects of bromodeoxyuridine and [3H]thymidine incorporation into DNA on cell proliferation, position, and fate. J Neurosci 31:15205–15217
4. Hoshino T, Nagashima T, Murovic J, Levin EM, Levin VA, Rupp SM (1985) Cell kinetic studies of in situ human brain tumors with bromodeoxyuridine. Cytometry 6: 627–632
5. Rakic P (2002) Neurogenesis in adult primate neocortex: an evaluation of the evidence. Nat Rev Neurosci 3:65–71
6. Nowakowski RS, Hayes NL (2000) New neurons: extraordinary evidence or extraordinary conclusion? Science 288:771
7. Kadar T, Horwitz V, Sahar R, Cohen M, Cohen L, Gez R et al (2011) Delayed loss of corneal epithelial stem cells in a chemical injury model associated with limbal stem cell deficiency in rabbits. Curr Eye Res 36:1098–1107
8. Chen W, Hara K, Tian Q, Zhao K, Yoshitomi T (2007) Existence of small slow-cycling Langerhans cells in the limbal basal epithelium that express ABCG2. Exp Eye Res 84: 626–634

9. Salic A, Mitchison TJ (2008) A chemical method for fast and sensitive detection of DNA synthesis in vivo. Proc Natl Acad Sci U S A 105:2415–2420
10. Buck SB, Bradford J, Gee KR, Agnew BJ, Clarke ST, Salic A (2008) Detection of S-phase cell cycle progression using 5-ethynyl-2′-deoxyuridine incorporation with click chemistry, an alternative to using 5-bromo-2′-deoxyuridine antibodies. Biotechniques 44:927–929
11. Diermeier-Daucher S, Clarke ST, Hill D, Vollmann-Zwerenz A, Bradford JA, Brockhoff G (2009) Cell type specific applicability of 5-ethynyl-2′-deoxyuridine (EdU) for dynamic proliferation assessment in flow cytometry. Cytometry A 75:535–546
12. Qu D, Wang G, Wang Z, Zhou L, Chi W, Cong S et al (2011) 5-Ethynyl-2′-deoxycytidine as a new agent for DNA labeling: detection of proliferating cells. Anal Biochem 417: 112–121
13. Gerdes J, Lemke H, Baisch H, Wacker HH, Schwab U, Stein H (1984) Cell cycle analysis of a cell proliferation-associated human nuclear antigen defined by the monoclonal antibody Ki-67. J Immunol 133:1710–1715

Chapter 5

Enrichment of Human Corneal Epithelial Stem/Progenitor Cells by Magnetic Bead Sorting Using SSEA4 as a Negative Marker

Martin N. Nakatsu and Sophie X. Deng

Abstract

The use of a specific antibody bound to magnetic beads to isolate subpopulations of cells is an efficient and simple technique that allows for the subsequent study of different cell populations. One important use of this isolation technique is the purification of stem cells from a mixed cell population. In this protocol, we describe a method to purify human corneal epithelial stem/progenitor cells or limbal stem cells (LSC), using stage-specific embryonic antigen-4 (SSEA4) as a negative surface marker.

Key words Cornea, Limbal stem cells, Limbal stem cell niche, Stage-specific embryonic antigen-4, Corneal epithelial stem/progenitor cells

1 Introduction

Magnetic bead cell sorting is a widely used method to efficiently separate pure subpopulations of cells [1], particularly in the isolation and enrichment of various types of stem cells. Several cell surface markers have been identified on embryonic stem cells (ESC) in human and mouse [2–5]. One of these ESC markers is SSEA4 [6], a globo-series carbohydrate core structure of glycoproteins [7]. SSEA4 has been previously used to isolate mesenchymal stem cells [8].

Interestingly, we have observed expression of SSEA4 in discrete clusters of the basal epithelial cells in the human limbus [9], where human corneal epithelial stem/progenitor cells are presumed to reside [10]. Contrary to a high expression level in ESC, SSEA4 is expressed at a much lower level in a subpopulation of limbal basal epithelial cells, but at a very high level in differentiated corneal epithelial cells. Further characterization of SSEA4 positive (+) and SSEA4 negative (−) limbal epithelial cells (LEC) shows that the

Bernice Wright and Che J. Connon (eds.), *Corneal Regenerative Medicine: Methods and Protocols*, Methods in Molecular Biology, vol. 1014, DOI 10.1007/978-1-62703-432-6_5, © Springer Science+Business Media New York 2013

SSEA4$^-$ population contained five times more cells of ≤11 μm in diameter (small cells) than the SSEA4$^+$ population.

The expression levels of putative LSC markers, ATP-binding cassette subfamily G member 2, ΔNp63α, N-cadherin and cytokeratin 14 were significantly higher in the SSEA4$^-$ population than in the SSEA4$^+$ population. SSEA4$^-$ cells also expressed a significantly lower level of the differentiation marker, cytokeratin 12. The colony-forming efficiency in the SSEA4$^-$ population was 1.6-fold ($P<0.05$) higher than in the SSEA4$^+$ population [9]. These findings suggest that SSEA4$^-$ LEC contain a higher proportion of limbal stem/progenitor cells and that SSEA4 can be used as a negative marker to enrich the stem/progenitor population of LEC.

Here, we describe a straightforward and efficient method to enrich corneal epithelial stem/progenitor cells from freshly isolated primary LEC using SSEA4 as a surface marker by using magnetic bead sorting. This method can be easily modified for the separation of stem/progenitor cell populations based on different cell surface markers and for further characterization of sorted LEC.

2 Materials

Perform all cell culture and isolation methods in a class II biological safety cabinet. Observe sterile techniques when working with LEC and magnetic beads. Prepare all reagents at room temperature (unless indicated otherwise). Prepare all cell culture media at 37 °C (unless indicated otherwise). Dispose of all biohazardous waste according to proper waste disposal regulations.

2.1 Magnetic Bead Sorting

1. For more detailed information regarding magnetic beads, see the product information on the Invitrogen website.
2. Dynabeads® SSEA4 kit (Invitrogen, Carlsbad, CA) includes the anti-SSEA4 antibody and the Depletion MyOne Streptavidin-coupled Dynabeads.
3. Isolation Buffer: calcium and magnesium-free phosphate buffered saline solution supplemented with 0.1 % (w/v) bovine serum albumin and 2 mM ethylenediaminetetraacetic acid (EDTA).
4. Rotary mixer (*see* **Note 1**).
5. Magnetic separation rack to hold 15 mL conical tubes and 1.5 mL microcentrifuge tubes.

2.2 Cell Culture

1. Growth Medium: Supplementary hormonal epithelial medium which consists of Dulbecco's Modified Eagle Medium/F12 supplemented with N2 (Invitrogen), 0.5 % (v/v) dimethyl sulfoxide, 2 ng/mL epidermal growth factor (EGF), 0.5 μg/mL hydrocortisone, 8.4 ng/mL cholera toxin A subunit, 5 % (v/v)

FBS, 100 U/mL penicillin, 100 μg/mL streptomycin, 10 μg/mL gentamicin and 0.25 μg/mL amphotericin B. An alternative growth medium, supplemented keratinocyte serum-free medium (KSFM), can also be used (*see* **Note 2**).

2. Dispase II.
3. 0.25 % (w/v) Trypsin–EDTA.
4. Microcentrifuge.
5. Hemocytometer.
6. Appropriate tissue culture apparatuses, including autoclaved/sterile surgical instruments, polyester-tipped applicators, 15 mL conical tubes, 1.5 mL microcentrifuge tubes, 2 mL microcentrifuge screw cap tubes and 60 mm diameter cell culture petri dishes.

3 Methods

Carry out all procedures in a class II biological safety cabinet at room temperature unless otherwise specified.

3.1 Washing Magnetic Beads

1. Resuspend Dynabeads® evenly in the original vial using a pipette.
2. Transfer 50 μL of the beads to a 1.5 mL microcentrifuge tube.
3. Mix the beads slowly several times with 1 mL of Isolation Buffer (*see* **Note 3**).
4. Place the tube containing beads on the magnetic separation rack for 1 min and then aspirate the supernatant (*see* **Note 4**).
5. Remove the tube from the magnetic separation rack and resuspend the washed beads in 50 μL of Isolation Buffer in preparation for the isolation procedure.

3.2 Cell Isolation

Numerous methods on the isolation of LEC have been published. We have had great success with our own protocol that was adapted from a previously described protocol [9, 11].

3.2.1 Isolation of Human Limbal Epithelial Cells

1. Punch out the central cornea from corneoscleral tissue using a 9 or 10 mm diameter trephine.
2. Remove the residual iris tissue, endothelium, conjunctiva and Tenon's capsule from the corneoscleral rim using surgical scissors.
3. Remove excess scleral tissue, leaving the 2 mm diameter limbal rim.
4. Incubate the limbal tissue in 2.4 U/mL sterile-filtered Dispase II in growth medium for 2 h.
5. Separate the LEC sheet from the stroma using surgical forceps.
6. Collect the LEC sheets in a 1.5 mL microcentrifuge tube.

7. Centrifuge cells at 300 × *g* for 5 min.
8. Aspirate the supernatant from the cell pellet.
9. Resuspend the cell pellet in 1 mL 0.25 % (w/v) Trypsin/EDTA and incubate for 10 min at 37 °C until a single-cell suspension is obtained (*see* **Note 5**).
10. Stop the enzymatic activity by adding 9 mL of growth medium.
11. Count cells using a hemocytometer (*see* Chapter 9, Subheading 3.1, **Step 9**).
12. Repeat **steps 7–8**.
13. Resuspend the LEC in growth medium at a density of 1×10^6 cells/mL in preparation for the cell sorting procedure (*see* **Note 6**).

3.2.2 Sorting of SSEA4+ and SSEA4− Limbal Epithelial Cells

1. Transfer 500 μL (0.5×10^6) of LEC suspended in growth medium to a 1.5 mL microcentrifuge tube.
2. Add 25 μL of anti-SSEA4 antibody to the LEC and mix slowly by gentle pipetting several times (*see* **Note 3**).
3. Incubate LEC/anti-SSEA4 antibody solution for 15 min at room temperature (*see* **Note** 7).
4. Add 1 mL of growth medium to the mixture and spin down the LEC for 8 min at 300 × *g*.
5. Aspirate the supernatant and resuspend the cell pellet in 500 μL of growth medium.
6. Add 50 μL of pre-washed Dynabeads® to the LEC.
7. Mix the LEC and magnetic beads slowly by gentle pipetting several times (*see* **Note 3**).
8. Incubate the mixture for 15 min at room temperature on a rotary mixer (*see* **Note 1**).
9. Add 1 mL of growth medium to the mixture and resuspend the bead-bound LEC by gentle pipetting several times (*see* **Note 3**).
10. Place the microcentrifuge tube containing the mixture of LEC and magnetic beads onto the magnetic separation rack for 2 min (*see* **Note 4**).
11. Collect the supernatant containing the bead-free SSEA4− LEC and transfer the cell supernatant to a clean 1.5 mL microcentrifuge tube.
12. Repeat **steps 9–11** to collect residual beads (*see* **Note 8**).
13. Resuspend bead-bound LEC containing the SSEA4+ population in 500 μL of growth medium.
14. Count cells using a hemocytometer (*see* **Note 9**).
15. The SSEA+ and SSEA− cell populations are now ready for phenotypic analysis.

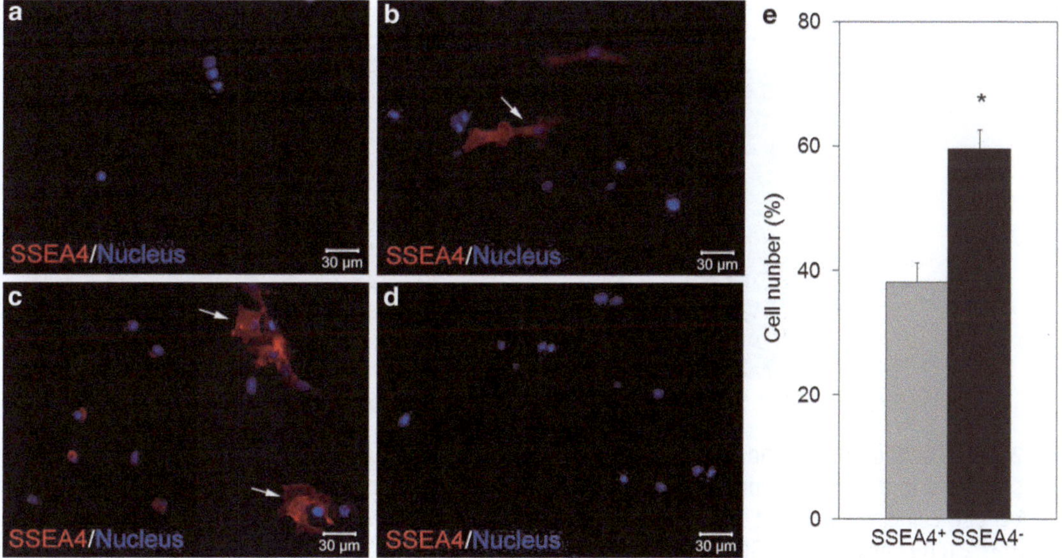

Fig. 1 Separation of SSEA4$^+$ and SSEA4$^-$ cell populations. (**a**) Unsorted LECs labeled with only a secondary antibody (negative control). (**b**) Unsorted cells contained both SSEA4$^+$ and SSEA4$^-$ populations. After sorting (**c**) the SSEA4$^+$ cell population was positive for SSEA4, while (**d**) the SSEA4$^-$ cell population was negative for SSEA4. (**e**) Quantitation of SSEA4 cell populations yielded 59.5 ± 5.7 % of SSEA4$^-$ cells and 38.1 ± 3.1 % of SSEA4$^+$ cells from the total LEC population. *P < 0.05

3.3 Cell Phenotype Analysis

1. To monitor the isolation efficiency of the SSEA4$^+$ and SSEA4$^-$ LEC populations, both cell populations can be immunolabeled with an anti-SSEA4 antibody (R&D Systems, Minneapolis, MN) (Fig. 1) (*see* **Note 10**). The separation efficiency should be >99 %.
2. Culture SSEA4$^+$ and SSEA4$^-$ LECs on growth-arrested mouse 3T3 fibroblasts for 14–21 days. Change the growth medium every 2–3 days (*see* **Note 11**).
3. Analyze the colony-forming efficiency (CFE) (*see* Chapter 3) of limbal stem/progenitor cells as previously described [9, 11] (Fig. 2) (*see* **Note 12**).

4 Notes

1. The use of a rotary mixer that rotates and tilts in a 360° motion is preferred.
2. Alternatively, keratinocyte serum-free medium supplemented with 10 % (v/v) FBS, 2 ng/mL EGF, 50 μg/mL bovine pituitary extract, 5 ng/mL fibroblast growth factor, 100 U/mL penicillin, 100 μg/mL streptomycin, 10 μg/mL gentamicin, and 0.25 μg/mL amphotericin B can be used as growth medium for the culture of LEC.

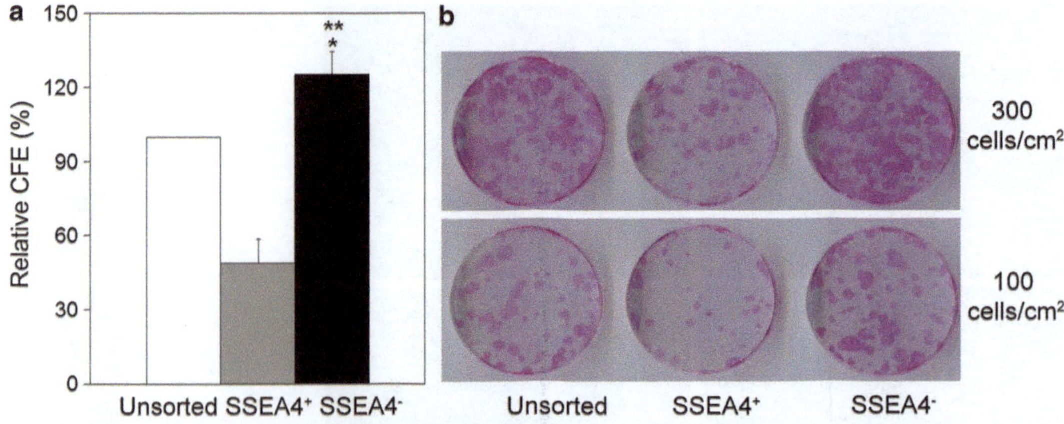

Fig. 2 CFE of sorted SSEA4$^+$ and SSEA4$^-$ cell populations. (**a**) The SSEA4$^-$ cell population had a much higher CFE compared to the SSEA4$^+$ and unsorted populations. (**b**) A representative image of unsorted, SSEA4$^+$ and SSEA4$^-$ LEC colonies seeded at 100 and 300 cells/cm^2. **P*<0.05 between the SSEA4$^-$ and SSEA4$^+$ populations. ***P*<0.05 between the SSEA4$^-$ and unsorted populations

3. Proper mixing of magnetic beads throughout the protocol is critical for the proper washing and binding of beads to the antibodies and cells. Pre-coat tubes and pipette tips with growth medium prior to mixing cells with beads. This will reduce the loss of cells/beads during transfer and washing.
4. Lean the magnetic rack at a 45° angle for more efficient collection of magnetic beads along the side of the microcentrifuge tube.
5. It is very important that LEC are in a single cell suspension and not in clumps.
6. LEC should always be resuspended in growth medium during the entire protocol to preserve viability.
7. The SSEA4 antibody is biotinylated and once the antibodies are bound to the SSEA4$^+$ LEC, the streptavidin-coated magnetic beads will bind to the biotinylated antibody-LEC complex.
8. Carefully check the SSEA4$^-$ LEC supernatant after repeating **steps 9–11** to make sure that there are no more residual beads left in the medium (*see* Subheading 3.2.2, **step 12**). Repeat **steps 9–11** again if the beads are still present.
9. The beads have no effect in cell counting or other downstream applications. We did not observe any issue in the isolation of RNA or protein from the bead-bound SSEA4$^+$ cell population.
10. For more detailed immunostaining methods, *see* ref. 9. We use the primary anti-SSEA4 antibody at a 1:10 dilution.

11. Mouse 3T3 fibroblasts can be growth arrested by mitomycin-C treatment or irradiation. We seed LEC at a cell density of 100 or 300 cells/cm^2.

12. For more detailed clonogenic assay methods, *see* refs. 9, 11 (*see* Chapter 3). We label limbal stem/progenitor cell colonies with Rhodamine B.

Acknowledgments

This work was supported by the Emily G. Plumb Estate and Trust, the California Institute for Regenerative Medicine (CIRM) TR2-01768 and the National Eye Institute (NEI) 5T32EY007026-35 (M.N.N.). All figures appearing in this protocol are the exclusive property of the Association for Research in Vision and Ophthalmology and are protected under the United States and International Copyright laws. The figures were taken from Truong et al. [9].

References

1. Owen CS, Sykes NL (1984) Magnetic labeling and cell sorting. J Immunol Methods 73: 41–48
2. Chambers I, Colby D, Robertson M, Nichols J, Lee S, Tweedie S et al (2003) Functional expression cloning of Nanog, a pluripotency sustaining factor in embryonic stem cells. Cell 113:643–655
3. Mitsui K, Tokuzawa Y, Itoh H, Segawa K, Murakami M, Takahashi K et al (2003) The homeoprotein Nanog is required for maintenance of pluripotency in mouse epiblast and ES cells. Cell 113:631–642
4. Adewumi O, Aflatoonian B, Ahrlund-Richter L, Amit M, Andrews PW, Beighton G et al (2007) Characterization of human embryonic stem cell lines by the International Stem Cell Initiative. Nat Biotechnol 25: 803–816
5. Niwa H, Miyazaki J, Smith AG (2000) Quantitative expression of Oct-3/4 defines differentiation, dedifferentiation or self-renewal of ES cells. Nat Genet 24:372–376
6. Draper JS, Pigott C, Thomson JA, Andrews PW (2002) Surface antigens of human embryonic stem cells: changes upon differentiation in culture. J Anat 200:249–258
7. Kannagi R, Cochran NA, Ishigami F, Hakomori S, Andrews PW, Knowles BB et al (1983) Stage-specific embryonic antigens (SSEA-3 and -4) are epitopes of a unique globo-series ganglioside isolated from human teratocarcinoma cells. EMBO J 2:2355–2361
8. Gang EJ, Bosnakovski D, Figueiredo CA, Visser JW, Perlingeiro RC (2007) SSEA-4 identifies mesenchymal stem cells from bone marrow. Blood 109:1743–1751
9. Truong TT, Huynh K, Nakatsu MN, Deng SX (2011) SSEA4 is a potential negative marker for the enrichment of human corneal epithelial stem/progenitor cells. Invest Ophthalmol Vis Sci 52:6315–6320
10. Schlotzer-Schrehardt U, Kruse FE (2005) Identification and characterization of limbal stem cells. Exp Eye Res 81:247–264
11. Nakatsu MN, Ding Z, Ng MY, Truong TT, Yu F, Deng SX (2011) Wnt/beta-catenin signaling regulates proliferation of human cornea epithelial stem/progenitor cells. Invest Ophthalmol Vis Sci 52:4734–4741

11. Mouse 3T3 fibroblasts can be growth arrested by mitomycin C treatment or irradiation. We seed 3T3s at a cell density of 100 or 500 cells/cm.

12. For more detailed clonogenic assay methods, see ref. 8 (also Chapter 3). We label human stem/progenitor cell colonies with Rhodamine B.

Acknowledgements

This work was supported by the Emily G. Plumb Estate and Trust, the California Institute for Regenerative Medicine (CIRM) [illegible] and the National Eye Institute (NEI) 5 R01EY0[illegible] [illegible]

Chapter 6

Limbal Epithelial Stem Cell Identification Using Immunoblotting Analysis

Bernice Wright and Che J. Connon

Abstract

The unambiguous identification of limbal epithelial stem cells is currently a major challenge in corneal stem cell biology. Specific molecular markers which characterize these cells are lacking. At present, the best strategy for identification of limbal epithelial stem cells is to investigate a variety of putative markers for these cells in a differentiated (cytokeratin (CK) 3: CK3, integrin $\alpha6$), undifferentiated (CK14), and naive state ($\Delta Np63\alpha$, ATP-binding cassette subfamily G member 2 (ABCG2), integrin $\alpha9$, Notch-1), alongside functional assays which indicate their stemness. The focus of this chapter is to highlight advances in the Western blotting technique for quantitative assessment of corneal epithelial cell markers, and the use of this technique for investigation of a range of different protein markers which identify limbal epithelial stem cells.

Key words Limbal epithelial stem cells, Immunoblotting, Limbal epithelial stem cell markers, Western blotting, Quantitative Western blotting

1 Introduction

Adult stem cells reside in the outer limbal region of the cornea (Fig. 1), widely accepted as the palisades of Vogt [1, 2]. Emerging evidence suggests, however, that these cells are also present in limbal crypts [3]. Furthermore, their exclusive location in the limbal region of the cornea has been questioned [4].

Alongside the controversy of their specific location in the cornea, a prevailing problem in the study of limbal epithelial stem cells (LESC) is the lack of a definitive identifying marker [5]. The detection of a range of cytokeratins in different regions of the cornea, by various methods including Western blotting, was previously reported to suggest the limbal location of epithelial stem cells [6]. A monoclonal antibody highly specific for a 64 kDa keratin, a marker for the advanced stages of corneal epithelial differentiation, demonstrated the presence of this marker in all layers of the cornea, but only in the suprabasal layers of the limbus. These reported

Bernice Wright and Che J. Connon (eds.), *Corneal Regenerative Medicine: Methods and Protocols*, Methods in Molecular Biology, vol. 1014, DOI 10.1007/978-1-62703-432-6_6, © Springer Science+Business Media New York 2013

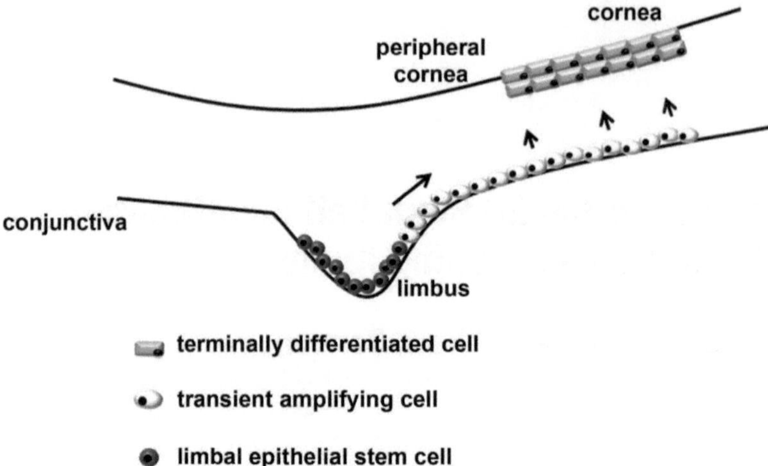

Fig. 1 The limbal stem cell niche. Corneal stem cells reside in the limbus at the corneoscleral junction between the conjunctiva and the cornea. Limbal epithelial stem cells regenerate the corneal surface by differentiating into transient amplifying epithelial cells which are present in the basal epithelia of the limbus and the peripheral cornea. Transient amplifying cells transform to terminally differentiated cells which reside within the suprabasal and superficial layers of the cornea

data indicated that less differentiated cells (i.e., stem cells) were present in the basal layer of the limbus and not present in the cornea. This study exemplified the need for markers to demonstrate the location of LESC.

A range of proteins have been suggested as candidates for the identification of corneal epithelial stem cells. These include CK14, CK19, ΔNp63α, ABCG2, integrin α9, integrin α6, integrin β1, CD71, enolase α, and the epidermal growth factor receptor [7]. More recent reports have suggested N-cadherin [8], Wnt/β catenin signalling [9], stage-specific embryonic antigen-4 (SSEA4) (negative marker) (*see* Chapter 5) [10], Notch-1 [11], and octamer-binding transcription factor 4 [12] as putative markers for corneal epithelial stem cells. Some markers, for example, CK14, have been demonstrated as unreliable due to their presence in both the cornea and the limbus [13]. Other markers including integrin α9, ΔNp63α, and ABCG2 are accepted as good indicators of the stemness of corneal epithelial cells [7]. The expression of markers for corneal epithelial cell differentiation (CK3, involucrin, connexin 43, integrin α6, E-cadherin, nestin) is often investigated alongside putative stem markers of these cells [7]. Reported evidence indicates that the best approach at present for identification of these cells is the investigation of a range of different markers (Fig. 2), the presence or the absence of which can be used collectively to make a consensus decision about the stemness of a population of limbal epithelial cells (LEC).

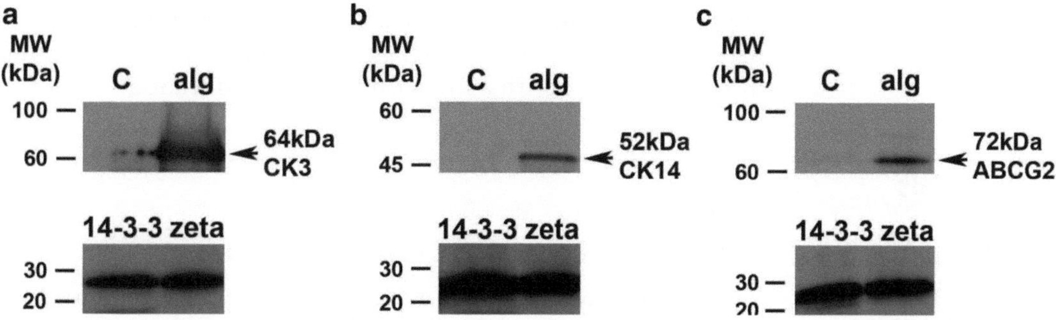

Fig. 2 Protein markers of differentiated (CK3: A), undifferentiated (CK14: B), and epithelial stem (ABCG2: C) cells. Freshly isolated (C) and alginate gel-encapsulated (alg) limbal epithelial cells (bovine) were lysed with ice-cold 1 % NP40 supplemented with protease and phosphatase inhibitors. Cell lysates were suspended in Laemmli buffer and proteins separated by SDS-PAGE, before immunodetection of CK3 (1/4,000 dilution: Chemicon), CK14 (1/8,000 dilution: Progen), and ABCG2 (1/500 dilution: New England Biolabs). Equivalent protein loading was verified by reprobing for 14-3-3 zeta. Blots represent three different experiments from three different corneoscleral buttons

Limbal epithelial stem cells were one of the first progenitor cell types to be used successfully for regenerative medicine. Currently, these cells are routinely used as therapy for limbal stem cell deficiency (LSCD) [14, 15]. Methodologies for the transplantation of LESC to treat LSCD are diverse. The main treatment involves the use of the biological substrate amniotic membrane (AM) for ex vivo expansion of these cells before transplantation [16, 17]. The therapeutic delivery of LESC is advancing to include the use of biomaterials [18] (*see* Chapters 11 and 12). LESC are also transplanted directly to treat LSCD [19], and important work is ongoing to develop methods to preserve the limited supply of these cells to improve their therapeutic use [20]. Therefore, it is vital to develop a toolbox of robust and reliable techniques for indisputable identification of LESC.

Amongst many of the techniques available to identify corneal epithelial stem cells (colony forming efficiency (CFE): *see* Chapter 3, bromodeoxyuridine (BrdU) labelling (*see* Chapter 4), quantitative PCR), Western blotting is one of the most robust. Western blotting (or protein blotting or immunoblotting) is a sensitive and powerful technique for the identification of proteins of a wide range of molecular masses and expression levels (particularly low-abundance proteins) [21]. This technique involves the transfer of proteins from a sodium dodecyl sulfate (SDS)-polyacrylamide gel to an adsorbent membrane [21]. Major advantages of Western blotting are that proteins immobilized on the membrane are equally accessible to different ligands, the same protein transfer can be used for multiple successive analyses, and the transfer process is efficient and relatively quick, so there is potential for high-throughput generation of data [21]. Western blotting is constantly evolving to allow more quantitative and sensitive detection of proteins. This chapter describes the basics of this technique and highlights advances in its use with a focus on LESC identification.

2 Materials

2.1 Preparation of Protein Samples from Corneal Cells

1. Isolated corneal cells (*see* Chapters 7 and 8).
2. Cell lysis buffer: A wide range of detergent-based buffers are available for lysing cells. Typically, cell lysis buffers contain Tris–HCl, ethylenediaminetetraacetic acid (EDTA), SDS, deoxycholate, Triton X, and/or nonidet P40. Commercially available cell lysis buffers include radio-immunoprecipitation assay (RIPA) buffer and CelLytic™ MT Cell Lysis Reagent (Sigma-Aldrich, St. Louis, MO, USA). The suitability of these buffers is dependent on the compatibility of the detergent with the assay used to estimate protein levels in samples (*see* **Note 1**).

   We use an NP40-based cell lysis buffer, as this nonionic detergent is compatible with our protein assay (the modified Lowry assay). The NP40 buffer we use comprises 2 % (v/v) NP40 (final concentration NP40: 1 % (v/v)), 300 mM NaCl, 20 mM Tris, and 10 mM EDTA at pH 7.3. We supplement this buffer with the serine protease inhibitors, PMSF (1 mM) and aprotinin (10 μg/mL), the serine/cysteine protease inhibitor, leupeptin (10 μg/mL), the tyrosine phosphatase inhibitor, sodium orthovanadate (2 mM), and the acid protease inhibitor, pepstatin A (1 μg/mL) (*see* **Note 2**).
3. Protein assay: There are a number of kits available to assess levels of total protein from cell lysates or protein solutions. Established protein assays include the Bradford assay, the Lowry assay, and the bicinchoninic acid (BCA) assay, which are available from Bio-Rad (Hercules, CA, USA), Thermo Scientific (Pierce: Rockford, IL, USA), and GE Healthcare (Buckinghamshire, UK).
4. UV/VIS spectrophotometer.
5. Optical cuvettes.
6. Laemmli sample treatment buffer (STB: 2× concentrate): 50 mM Tris–HCl, 4 % (w/v) SDS, 10 % (v/v) 2-mercapto-ethanol (can be omitted to run samples under nonreducing conditions (*see* **Note 3**)), 20 % (v/v) glycerol, a trace of bromophenol blue (or Coomassie brilliant blue R), pH 6.8.

2.2 Preparation of Acrylamide Gels

1. Sodium dodecyl sulfate-polyacrylamide gel electrophoresis (SDS-PAGE) system (mini or maxi system) and electrophoresis power supply. SDS-PAGE systems and appropriate power supplies are available from Bio-Rad and GE Healthcare.

2. 30 % (w/v) acrylamide:0.8 % (w/v) bisacrylamide solution. A commercially available premade acrylamide solution (Protogel: National Diagnostics: Hull, UK) is recommended to minimize handling of highly toxic acrylamide solutions.
3. Resolving gel buffer: 3 M Tris–HCl, pH 8.8.
4. Stacking gel buffer: 0.5 M Tris–HCl, pH 6.8.
5. 10 % (w/v) SDS solution.
6. 1.5 % (w/v) ammonium persulfate solution (prepared immediately before use).
7. *N*,*N*,*N*′,*N*′-tetramethylethylenediamine (TEMED).
8. 50 % (v/v) butan-1-ol. Prepare by mixing equal proportions of water and butan-1-ol, allow phases to separate, and use the upper phase.
9. Tris–SDS–glycine gel running buffer. 10× concentrate: 250 mM Tris–HCl, 1.92 M glycine, 1 % (w/v) SDS, pH 8.3. Dilute 10-fold before use. This buffer is available from National Diagnostics and Fisher Scientific.
10. 70 % (v/v) Ethanol.

2.3 Preparation of Gradient Gels

1. Linear gradient former (suppliers: Bio-Rad and GE Healthcare).
2. Peristaltic pump and tubing (Watson-Marlow Pumps Group: Cornwall, UK).
3. Glycerol.

2.4 Loading and Running Gels

1. Heating block or boiling water bath to heat samples to 100 °C.
2. Pre-stained broad-range molecular weight markers. Precision Plus markers from Bio-Rad and Colorburst markers from Sigma are suitable.
3. Gel-loading tips (suppliers: Starlabs (Milton Keynes, UK), Fisher Scientific, SLS, Greiner Bio-one).

2.5 Semidry Western Blotting

1. Western blot electrotransfer cell (e.g., Trans Blot SD, Bio-Rad) and appropriate power supply (can be obtained together with blotting systems from Bio-Rad, Fisher Scientific, or GE Healthcare).
2. Whatman 3 MM chromatography paper (Fisher Scientific) cut to the dimensions of the gel to be blotted.
3. Polyvinylidine difluoride (PVDF) or nitrocellulose membranes (suppliers: Bio-Rad, GE healthcare, Fisher Scientific).
4. Anode buffer I: 0.3 M Tris–HCl, 20 % (v/v) methanol pH 10.4.
5. Anode buffer II: 25 mM Tris–HCl, 20 % (v/v) methanol pH 10.4.

6. Cathode buffer: 25 mM Tris–HCl, 40 mM 6-amino-n-caproic acid (Sigma), 20 % (v/v) methanol pH 9.2.
7. Large test tube.
8. 100 % analytical grade methanol.

2.6 Wet Western Blotting

1. Wet Western blotting system (e.g., Mini Trans-Blot Cell, Bio-Rad) and appropriate power supply.
2. Large tray for blotter cell assembly submerged in buffer.
3. CAPS transfer buffer: 10 mM 3-[cyclohexylamino]-1-propanesulfonic acid (CAPS: Sigma), 10 % (v/v) methanol, pH 11.
4. PVDF or nitrocellulose membranes.
5. 100 % analytical grade methanol.

2.7 Coomassie Staining of Gels After Protein Transfer

1. Coomassie brilliant blue R stain solution: 50 % (v/v) methanol, 10 % (v/v) glacial acetic acid, 0.1 % (w/v) Coomassie brilliant blue R (Sigma), prepared in deionized water.
2. Coomassie destaining solution: 50 % (v/v) methanol, 10 % (v/v) glacial acetic acid, prepared in deionized water.
3. Staining trays.
4. 100 % analytical grade methanol.

2.8 Ponceau Staining of Membranes After Protein Transfer

1. Ponceau stain solution: 0.1 % (w/v) Ponceau S, 5 % (v/v) glacial acetic acid prepared in deionized water.
2. Deionized water.

2.9 Visualizing Proteins on Membranes

1. Bovine serum albumin (BSA) or powdered milk for blocking nonspecific binding of antibody probes to the membrane.
2. Tris-buffered saline–Tween 20 (TBS-T) buffer: Tris base, NaCl, Tween 20 pH 7.6
3. Phosphate-buffered saline (PBS).
4. Primary antibodies: We typically use p63 (Millipore, Chemicon: Watford, UK), ABCG2 (Millipore, Abcam: Cambridge, UK and Cell Signalling Technology: Boston, MA, USA), CK3 (Millipore), CK14 (Progen Scientific: Merton, UK), integrin $\alpha 6$ (Cell Signalling Technology), and integrin $\alpha 9$ (Cell Signalling Technology) as LESC and LEC markers.
5. Horse radish peroxidase (HRP)-conjugated secondary antibodies. (suppliers: Cell Signalling Technology, R and D systems: Abingdon, UK; GE Healthcare; Santa Cruz, Autogen Bioclear: Wiltshire, UK; and Jackson Laboratories, Stratech: Suffolk, UK).
6. Containers for washing membranes.
7. Polyethylene flat tubing for incubating blots with antibodies.
8. Saran wrap.

9. Enhanced chemiluminescence (ECL) system (suppliers: Pierce, GE Healthcare).
10. X ray film (PLH medical: Hertfordshire, UK), film developing equipment and reagents (Agfa-Gevaert: Mortsel, Belgium).
11. X ray film cassettes (Kodak: Michigan, USA).
12. Gel/blot documentation system: A number of gel and blot documentation/imaging systems are available. At the top of the range is the Typhoon™ 9400 (GE Healthcare). Other systems include the ChemiDoc™ MP system and the Gel Doc™ XR+ and XRS+ Imagers from Bio-Rad, the FluorChem E and M Imagers (Protein Simple: Hertfordshire, UK), and the G:Box range (Syngene: Cambridge, UK).

2.10 Stripping Primary Antibodies from Membranes

1. Stripping buffer: 2 % (w/v) SDS (Sigma), 5 % (w/v) β-mercaptoethanol (Sigma) in 1× TBS-T. Stripping buffers are commercially available (Pierce, Li-Cor: Cambridge, UK, Amresco: Solon, OH, USA).
2. 1× TBS-T.
3. Deionized water.
4. Water bath heated to 80 °C.
5. 2.5, 5, or 10 % (w/v) BSA or milk dissolved in 1× TBS-T.
6. Secondary antibodies: After the stripping process, we typically reprobe either the membrane with an antibody used to verify equal loading (anti-14-3-3 zeta (Santa Cruz), anti-GAPDH (Abcam), anti-tubulin (Sigma), or anti-actin (Santa Cruz) antibodies) or an antibody against a different LESC marker.

2.11 Coomassie and Silver Staining of Membranes

1. Coomassie brilliant blue G colloidal stain concentrate: 2 % (v/v) phosphoric acid, 15 % (w/v) ammonium sulfate, 0.1 % (w/v) Coomassie brilliant blue G (Sigma) prepared in deionized water.
2. 100 % analytical grade methanol.
3. Staining trays.
4. Silver nitrate stain solution: 40 % (w/v) sodium citrate (Sigma), 20 % (w/v) ferric sulfate (Sigma), 20 % (w/v) silver nitrate (Sigma).
5. Silver destaining solution: 30 % (w/v) potassium ferricyanide (Sigma), 10 % (w/v) sodium carbonate (Sigma), and 60 % (w/v) sodium thiosulfate (Fisher Scientific).
6. 5 % (v/v) glacial acetic acid.
7. Deionized water.

3 Methods

3.1 Preparation of Protein Samples

The main principle of a protein assay is to estimate total protein levels in a cell lysate or a protein solution, using a standard curve derived from known concentrations of BSA or gamma globulin.

The total protein concentration is demonstrated by a color change of the sample solution in proportion to the protein concentration, which is measured using colorimetric techniques. We use a modified version of the Lowry protein assay. The optimal protein concentration range for the Lowry method is 0.01–1.0 mg/mL. This method is performed under alkaline conditions, and combines the reactions of copper ions and protein peptide bonds with oxidation of aromatic protein residues (the Biuret test). Copper produced by the oxidation of peptide bonds reacts with the Folin–Ciocalteu reagent (phosphotungstic acid and phosphomolybdic acid) [22]. The Folin reagent is reduced and aromatic protein residues (tyrosine (Tyr) and tryptophan (Trp)) are oxidized. The absorbance of the reduced Folin reagent is measured at 750 nm using a spectrophotometer, and the total protein concentration is expressed as the concentration of Trp and Tyr residues that reduce the Folin reagent [22].

1. For the modified Lowry protein assay, ensure that all reagents are at room temperature and the spectrophotometer is calibrated.
2. Prepare serial dilutions of gamma globulin or BSA standard.
3. For the modified Lowry protein assay, we use 10 μL of sample or standard, 100 μL of reagent A, and 800 μL of reagent B. We add these sequentially into the optical cuvette (i.e., sample first, reagent A second, and reagent B third). Three replicates of samples and standard are analyzed and an average absorbance value is obtained.
4. Measure the absorbance of the protein standards and samples at 750 nm using the calibrated spectrophotometer.
5. Use the average absorbance values of the standards to draw a curve (using Microsoft Excel or GraphPad Prism), and define the curve with an equation relating *x*-axis values (concentration (mg.mL-1)) to *y*-axis values (absorbance). As the absorbance of protein samples is known, this equation can be used to calculate total protein concentrations of samples.
6. Solubilize cell lysates in Laemmli buffer with or without a reducing agent (β-mercaptoethanol) (*see* **Note 3**). At this stage, protein samples will be ready for separation on the SDS-PAGE gel.

3.2 Preparation of Acrylamide Gels

SDS-PAGE is a technique involving the separation of proteins by their molecular masses by electrophoretic migration through a polyacrylamide gel. Polyacrylamide gels are molecular sieves formed through the polymerization of acrylamide monomers into long polymers cross-linked by bisacrylamide [23]. The polymerization process is aided by ammonium persulfate and TEMED. Ammonium persulfate is a source of free radicals which initiates gel formation and TEMED, the catalyst, stabilizes free radicals, and improves gel polymerization [23]. The porosity of the gel scaffold is dependent on acrylamide concentration, i.e., as acrylamide concentration increases, pore size decreases. Proteins with a net negative charge conferred by SDS can be separated through acrylamide gels by their size. As current is passed through the gel, proteins migrate towards the anode. Molecular weight markers are included to determine the apparent molecular mass of separated proteins.

A discontinuous acrylamide gel system (comprising a stacking gel and a resolving gel) that is suitable for the separation of most protein mixtures can be used. The percentage of acrylamide in the resolving gel can be varied to provide optimal sieving conditions for proteins of interest (Table 1). To separate high-molecular-mass proteins (100–250 kDa), low-percentage acrylamide gels (6 or 8 %) are suitable and for low-molecular-mass proteins (10–30 kDa), high-percentage acrylamide gels (12, 15, or 18 %) are required. For proteins of intermediate molecular mass (40–80 kDa), a 10 % acrylamide gel is usually sufficient. If, however, a number of proteins of varying molecular weights are investigated together, a gradient resolving gel that allows separation of a range of proteins of different sizes is the best option. Acrylamide should be handled with care, as this reagent is carcinogenic and neurotoxic.

1. Clean gel assembly apparatus before use. These include glass plates, sample well combs, gel extractors, gaskets, and stands. Wash apparatus with warm water and detergent before wiping thoroughly with 70 % (v/v) ethanol. Assemble the gel cassette by following the manufacturer's instructions. To check that the assembled gel cassette does not leak, fill this with water and check for leaks after 5–10 min (*see* **Note 4**).
2. Determine the depth of the stacking and resolving gels before casting. The entire depth of the stacking gel should be twice the length of the wells shaped into it. This distance should be measured on the glass plates and marked with a marker pen. The remaining lower half below the upper marked section will contain the resolving gel (*see* **Note 5**).
3. Prepare the resolving and stacking gel mixtures by combining all reagents (except ammonium persulfate and TEMED) in clean 50 mL falcon tubes or beakers (*see* Table 1). Check that the gel cassette is watertight before finishing preparation of the

Table 1
Gel preparation for acrylamide gels

	4 % (stacking gel)		8 %		10 %		12 %		15 %		18 %	
	Mini	**Maxi**	**Mini**	**Maxi**	**Mini**	**Maxi**	**Mini**	**Maxi**	**Mini**	**Maxi**	**Mini**	**Maxi**
Acrylamide:bisacrylamide (30:0.8 % [w/v]) (mL)	0.666	3.333	2.667	18.667	3.333	23.333	4.000	28.000	5.000	35.000	6.000	42.000
Resolving gel buffer (mL)	–	–	1.250	8.750	1.250	8.750	1.250	8.750	1.250	8.750	1.250	8.750
Stacking gel buffer (mL)	1.250	6.250	–	–	–	–	–	–	–	–	–	–
10 % (w/v) SDS (mL)	0.050	0.250	0.100	0.700	0.100	0.700	0.100	0.700	0.100	0.700	0.100	0.700
Water mL	2.780	13.897	5.478	38.348	4.812	33.682	4.145	29.015	3.145	22.015	2.145	15.015
1.5 % Ammonium persulfate (mL)	0.250	1.250	0.500	3.500	0.500	3.500	0.500	3.500	0.500	3.500	0.500	3.500
TEMED (mL)	0.004	0.020	0.005	0.035	0.005	0.035	0.005	0.035	0.005	0.035	0.005	0.035

Quantities are sufficient to prepare two mini or two maxi gels. Mini gel dimensions: Approximately 80 × 60 × 1 mm. Maxi gel dimensions: Approximately 160 × 120 × 1.5 mm resolving gel mixture by adding ammonium persulfate and TEMED. Swirl the resolving gel mixture gently to ensure even polymerization (*see* **Note 6**). Apparatus that have been in contact with unpolymerized acrylamide should be disposed of or cleaned appropriately.

4. Replace the water in the gel cassette with the resolving gel mixture using a pipette. Gently layer saturated butan-1-ol (2–3 mm depth) over the gel mixture in the cassette to smooth out the upper edge of the gel and prevent it from drying out after polymerization. Allow approximately 30 min for the gel to polymerize at room temperature (*see* **Note 6**).
5. Remove butan-1-ol from the gel cassette. Wash the inside and top of the glass plates with an excess of deionized water to wash out the butan-1-ol.
6. Add appropriate volumes of ammonium persulfate and TEMED to the stacking gel mixture. To ensure even polymerization, mix by inverting gently. Fill the remaining top section of the gel cassette with stacking gel mixture, taking care to avoid bubbles (use a pipette to remove any bubbles which form).
7. Lower the sample well comb into the stacking gel mixture gently to avoid forming bubbles. Excess gel mixture that overflows may be used to top up the gel.
8. The stacking gel will polymerize in approximately 30 min at room temperature. Gently remove the comb after the gel has polymerized and wash the wells with an excess of deionized water to remove partially polymerized acrylamide. Shake the gel assembly gently to remove water from the wells.
9. Assemble the gel cassettes with the electrodes to form reservoirs and insert into the tank. Add appropriate volumes of Tris–glycine–SDS gel running buffer to the reservoir and the tank, following the manufacturer's instructions.

3.3 Preparation of Gradient Gels

To separate proteins with a range of molecular masses, a gradient gel should be used. Gradient gels contain a gradually increasing percentage of acrylamide from the top of the gel to the bottom. This allows the separation of high-molecular-mass proteins at the top of the gel and low-molecular-mass proteins at the bottom of the gel. The gradient is formed by pouring two resolving gel solutions, each containing the desired percentage of acrylamide using a linear gradient former and a peristaltic pump. A 10–18 % gradient is standard, but 6–18 % or 8–18 % gradients may be used, depending on the sizes of proteins of interest.

1. Follow Table 2 to prepare gel mixtures, but do not add ammonium persulphate or TEMED until the gradient apparatus is assembled. The volumes of reagents can be modified as required to prepare gels of various dimensions.

Table 2
Gel preparation for a 10–18 % gradient gel

	4 % (stacking gel)	10 %	18 %
Acrylamide:bisacrylamide (30:0.8 % [w/v]) (mL)	1.667	5.833	10.500
Resolving gel buffer (mL)	–	2.190	2.190
Stacking gel buffer (mL)	3.125	–	–
10 % (w/v) SDS (mL)	0.125	0.175	0.175
Water (mL)	6.948	8.200	2.753
Glycerol (mL)	–	0.200	1.000
1.5 % Ammonium persulfate (mL)	0.625	0.875	0.875
TEMED (mL)	0.010	0.007	0.007

Quantities of reagents are sufficient to prepare a maxi gel (dimensions: approximately 160 × 120 × 1.5 mm). Quantities can be reduced as required to prepare mini gels

2. Assemble the gel cassette as described in Subheading 3.2, **step 1**, and place this apparatus together with the acrylamide solutions and gradient former in the cold room at 4 °C. This is to ensure that the gel does not polymerize as it is being poured (low temperatures slow down polymerization).
3. Connect the pump to the gradient former using peristaltic pump tubing with an internal diameter of approximately 1 mm. Attach a plastic pipette tip to the free end of the tubing and prize this between the glass plates. Adjust the pump to a flow rate of 5 mL/min.
4. Add ammonium persulfate and TEMED to the acrylamide solutions and swirl gently to mix. Transfer the higher percentage acrylamide gel mixture to the chamber of the gradient former from which the pump draws, and the lower percentage acrylamide gel mixture to the adjoining chamber. Open the tap in the chamber to allow flow between the chambers before turning on the pump.
5. Following transfer of acrylamide solutions into the gel cassette, layer water-saturated butan-1-ol (approximately 2–3 mm) over the gel. Avoid moving the gel at this stage, as the gradient and polymerization process will be disturbed (*see* **Note 6**). Cast the stacking gel as described in Subheading 3.2, **step 6**.

3.4 Running and Loading Gels

1. Heat protein samples solubilized in Laemmli buffer to 95 °C using a heating block or a boiling water bath for 5 min. Heating further denatures proteins ensuring that they are effectively coated in SDS and are fully.

2. Pipette samples into the bottom of the wells of the stacking gel using a gel-loading tip (*see* **Note 7**). To run all samples at the same rate, keep the volume in each well identical (add Laemmli sample treatment buffer to wells to achieve equivalent volumes). To prevent spillovers during loading, avoid mixing samples in the wells.
3. Attach the tank to a power supply and run minigels (dimensions) at 100 V (constant) per gel. The current will vary due to increased electrical resistance as the gels run. Therefore, either the current or the voltage should be fixed (refer to manufacturer's instructions) (*see* **Note 8**).
4. Turn off the power supply when the blue dye has been eluted from the gel cassette or has approached the bottom of the gel. To avoid diffusion of proteins, immediately begin the Western blotting step (*see* **Note 9**).

3.5 Semidry Western Blotting

Semidry Western blotting is suitable for the transfer of corneal proteins of a range of molecular masses. This system uses three different buffer solutions to aid protein transfer. To avoid transfer of exogenous protein to blots, apparatus should be cleaned thoroughly before use, and gloves should be worn during the preparation of the membrane and filter paper.

1. Soak four pieces of Whatman 3 MM chromatography paper in anode buffer I, two pieces in anode buffer II, and six pieces in cathode buffer. Soak PVDF membrane in analytical grade methanol (100 % (v/v)) or nitrocellulose in anode buffer II (*see* **Note 10**).
2. Clean the electroblotter surfaces with deionized water before placing the anode I buffer-soaked paper on the anode (Bio-Rad Transblot SD—bottom plate) of the blotter. Place the two sheets of anode II buffer on top of the other sheets and roll flat from the middle to the edges.
3. Place the PVDF or nitrocellulose membrane on top of the sheets, taking care to handle only by the edges (using forceps). Pipette some water onto the membrane to provide a wet surface for the gel.
4. Use the gel-extracting tool to open the gel cassette (pry glass plates apart), pick up the gel by the edges, and lower gently onto the membrane. The top of the gel can be flooded with cathode buffer to allow it to slide in place over the membrane and remain moist. Ensure that all bubbles are removed from under the gel by gently rolling a pipette over the gel or gently pushing them out with fingers.
5. Place the six pieces of paper soaked in cathode buffer over the gel carefully so that the gel is not moved. Align this with the other layers of paper and roll out any bubbles without placing too much pressure on the gel underneath.

6. Remove excess buffer from around the gel-membrane assembly, attach the cathode electrode, and plug the blotter into the power supply.
7. Transfer proteins at 15 V constant voltage for 2 h.
8. Remove the membrane from the transfer unit and cut a corner of the blot to determine its orientation in the future. Proteins transferred to a PVDF membrane will be on the surface that was in contact with the gel. The gel should be Coomassie-stained to ensure complete transfer of proteins.

3.6 Wet Western Blotting

The advantage of wet blotting against semidry blotting is that proteins can be transferred more quickly. The following protocol is for use with the Bio-Rad Mini Trans-Blot Cell.

1. Equilibrate the SDS-PAGE in CAPS transfer buffer for 20 min before transfer.
2. Assemble the transfer cassette in a tray. Place a piece of Whatman number 1 paper on each piece of padding (cut to the same dimensions as the padding), before filling the tray with CAPS buffer to the level of the padding.
3. Soak PVDF or nitrocellulose membrane in 100 % analytical grade methanol, before equilibrating in CAPS buffer for 5 min.
4. Place the gel on the filter paper lining the cathode side of the cassette (grey), ensuring that there are no bubbles trapped under the gel. Cover the gel in CAPS buffer to prevent it drying out.
5. Place the membrane over the gel, ensuring that no bubbles become trapped and cover this with the remaining piece of filter paper. Take care that the cassette is completely soaked in CAPS buffer, close this, and transfer the assembly to the tank.
6. Apply 200 mA of constant current for 20–30 min to transfer proteins.
7. Remove the membrane from the transfer unit and cut a corner of the blot to determine its orientation in the future. Proteins transferred to a PVDF membrane will be on the surface that was in contact with the gel. The gel should be Coomassie-stained to verify complete transfer of proteins.

3.7 Coomassie Staining of Gels

To validate the transfer of proteins from the gel to the membrane, the gel can be stained with Coomassie brilliant blue R, after the transfer process. This is an all-purpose stain; more sensitive stains which can detect lower quantities of protein include Coomassie brilliant blue G colloidal stain, silver stain, and the Deep Purple fluorescent stain (GE Healthcare).

1. After the transfer process, submerge the gel into the stain solution. Incubate with gentle agitation at room temperature for approximately 2 h or overnight.
2. Wash the gel gently with deionized water before placing in destain solution. The destain solution will become less transparent as the stain is removed from the gel; replace the destain solution as required until the background on the gel is clear and protein bands are visible.
3. The gel should be washed in deionized water and ideally imaged immediately. We find, however, that the gel can be stored in deionized water for up to 4 weeks or 5 % (v/v) acetic acid for up to 8 weeks.

3.8 Ponceau Staining of Membranes

For further validation of the transfer of proteins from the gel to the membrane, the membrane can be stained with Ponceau S, after the transfer process.

1. Submerge membrane in Ponceau stain solution for 5–10 min and observe for visible protein bands.
2. When protein bands become visible, remove the Ponceau S stain from the membrane by washing with deionized water and continue with the blotting process.

3.9 Visualizing Proteins on Membranes

Antibody probes used to identify proteins immobilized on membranes are conjugated to either fluorescent molecules (e.g., Cye dyes) or protein substrates (e.g., HRP). The HRP reporter is reacted with hydrogen peroxide, producing peroxide which reacts with luminol to generate 3-aminophthalate, the luminescent product. 3-aminophthalate emits light at 425 nm that can be captured on X-ray film or a gel/blot imaging system. Fluorescently labelled antibody probes are excited by light and the resulting emitted light is detected by a gel/blot imaging system photosensor. Typically, this is a CCD camera equipped with appropriate emission filters which capture a digital image of labelled proteins.

Fluorescent detection of proteins on Western blots is more quantitative than the chemiluminescent enzyme reaction system. The fluorescent signal is a result of the excitation of molecules leading to the transient release of photons before the molecule returns to its normal state. Enzyme–substrate systems produce a colored product or light emission due to a chemical reaction. The signal from individual photons provides a direct measure of the quantity of labelled proton, but the signal from enzyme–substrate systems is less linear. Fluorescent blotting is, however, less sensitive than chemiluminescence (see ref. 24 for a more thorough discussion about the advantages and limitations of fluorescent and chemiluminescent detection of proteins on Western blots).

1. Block nonspecific binding on the membrane with 2.5, 5, or 10 % (w/v) BSA (*see* **Note 11**) or milk, dissolved in 1× TBS-T. Incubate membranes for 1 h at room temperature with gentle agitation or overnight at 4 °C in blocking solution.
2. Seal blocked membranes in polyethylene bags and add antibodies diluted 2 % (w/v) BSA or milk to the protein-side of the membrane.
3. Incubate the membrane with primary antibodies at room temperature with rotation. The length of incubation depends on the antibody. We typically incubate blots with primary antibodies for 1–2 h at room temperature or overnight at 4 °C with rotation (*see* **Note 12**).
4. Wash the membrane with 1× TBS-T after incubating with the primary antibody. The number of washes can vary, e.g., seven 5 min washes, three 15 min washes, or two 1 h washes, depending on the optimization procedure for the antibody used (*see* **Note 13**).
5. Incubate the membrane with secondary antibodies at room temperature for 1–2 h (*see* **Note 12**).
6. Wash the membrane with 1× TBS-T. If the antibody probe is conjugated to a fluorescent tag, wash the membrane with 1× PBS before imaging (*see* **Note 14**).
7. Image the Western blot using X-ray film or a gel/blot imaging system.

3.10 X-Ray Film Imaging of Proteins on Membranes

1. Place the membrane, protein side up, on a large piece of saran wrap. Pipette the ECL reagent onto the membrane and incubate for 1 min (*see* **Note 15**). Turn the membrane protein side down onto another piece of saran wrap and enclose the membrane in the saran wrap. Attach the membrane (protein side up) into a film cassette and expose the X-ray film to luminescent proteins on the membrane in a darkroom. Avoid saturating the film; start with short 5 s exposure and then perform longer exposures as needed (*see* **Note 16**).
2. Use densitometry software (e.g., ImageQuant: GE Healthcare, Quantity One: Bio-Rad) to obtain semi-quantification of imaged protein bands on X-ray film.

3.11 Visualizing Proteins on Membranes Using Gel/Blot Imaging Systems

Place membranes (protein side down) directly onto the surface of the scanner, taking care to remove any bubbles. Adjust the voltage as required to obtain a clear unsaturated image of protein bands. The settings can be saved for subsequent experiments involving analysis of the same proteins and antibodies. Using identical setting for experimental replicates also reduces variation between different Western blots. Refer to the manufacturer's guidelines for the use of currently available imaging systems (Typhoon™ 9400 (GE Healthcare), Gel Doc™ XR Imager (Bio-Rad), and FluorChem E and M Imagers (Protein Simple)).

3.12 Stripping Antibodies from Membranes

Membranes can be repeatedly probed with different antibodies to investigate levels of different proteins on the same Western blot. An advantage of this procedure is that it allows comparison of the expression of several functionally related proteins simultaneously. Membranes are often stripped to investigate a protein which is used to normalize sample loading (a loading control), for example, GAPDH, actin, and tubulin. In cases involving analysis of phosphorylated proteins (i.e., phospho p63), the loading control antibody is raised against p63.

1. Heat the water bath and stripping buffer to 80 °C.
2. Incubate the membrane with the heated stripping buffer (submerge membrane in the buffer) in a sealed container. Balance the sealed container containing the membrane and buffer in the water bath (*see* **Note 17**).
3. Strip the blot for 20 min, agitating the container every 5 min to aid the stripping process.
4. In a fume hood, pour the stripping buffer away from the membrane for reuse (stripping buffer can be used up to three times) and wash the membrane with deionized water. Wash the membrane until the odor of β-mercaptoethanol is faint.
5. Wash the membrane for a final time in 1× TBS-T for 5 min with gentle agitation.
6. Block the membrane with BSA or milk for 1 h at room temperature or overnight at 4 °C before incubation with antibodies.

3.13 Coomassie and Silver Staining of Blots

The Coomassie brilliant blue G colloidal stain and silver stain can be used to detect proteins that are present on membranes at low levels that are difficult to detect using antibodies.

3.13.1 Coomassie Brilliant Blue G Staining

Prepare the stain by mixing four parts Coomassie brilliant blue G colloidal concentrate with one part 100 % methanol.

1. Incubate the blot with stain at room temperature with gentle agitation for several hours or overnight. For greater sensitivity the blot can be stained for several days (wrap the container in parafilm to reduce evaporation of the stain).
2. Remove background staining from the blot by washing for several hours in deionized water.
3. Allow the blot to air-dry at room temperature and protect in a plastic transparent wallet before imaging.

3.13.2 Silver Staining

1. Silver stain membranes in the silver solution at room temperature with gentle agitation for 20 min.
2. Wash the stained membranes with distilled water before removing background staining with destain solution.
3. Allow the blot to air-dry at room temperature and protect in a plastic transparent wallet before imaging.

4 Notes

1. Choose a detergent that is compatible with the assay used to determine protein levels. Detergents which are suitable for use with the Lowry assay include Triton X-100 (0.1 % (w/v)), SDS (0.1 % (w/v)), and NP40 (1 % (w/v)). The Bradford assay is compatible with only low concentrations of commonly used detergents, and the BCA assay is compatible with up to 5 % (w/v) NP40, SDS, Triton X-100, or Brij-35.
2. Protease and phosphatase inhibitors are added to cell lysis buffers to ensure that proteins of interest (and their activation states) are not degraded by endogenous proteolytic and phospholytic enzymes, released from subcellular compartments of lysed cells. Inhibitors degrade quickly, so they should be prepared as required. These inhibitors are available individually or as cocktail solutions from suppliers including Pierce, Sigma, Roche Applied Science (Burgess Hill, UK), and Merck (Middlesex, UK).
3. The sample treatment buffer (STB) we use includes the reducing agent, β-mercaptoethanol (dithiothreitol is an alternative), to denature proteins through disruption of disulfide bonds. Some proteins may only be detected under nonreducing conditions (in the absence of a reducing agent), depending on the specificity of the antibody used. In these instances, the reducing agent can be replaced with water in the STB.
4. To avoid leaks, check that the glass plates are not chipped and that the bottom of the spacers is exactly flush with the glass plates.
5. Cast gels can be made and stored overnight at 4 °C before use. To keep gels hydrated, they should be wrapped in cling film with a layer of gel running buffer or deionized water in the sample wells. Longer storage is not recommended. Precast gels are available from Bio-Rad and GE Healthcare.
6. Avoid moving the gel cassette to check for polymerization. An interface will appear showing excluded buffer at the top of the polymerized gel and below the layer of butan-1-ol. At this stage, allow a further 5–10 min to ensure complete

polymerization of the gel. Low temperatures (4 °C) slow down the polymerization process; therefore gradient gels will take longer than 2–3 h to polymerize. To ensure that the gradient is not disturbed, gradient gels should not be moved.

7. When loading samples onto the gel lower the gel loading pipette tip to the bottom of the sample well and slowly draw this up as the sample is extruded into the well. The glycerol in the STB will weight the sample to the bottom of the well, and the bromophenol blue/Coomassie dye will make the sample visible.
8. To ensure efficient stacking of proteins, the first part of the gel should be run slowly, at a low voltage. Gels can be run at the recommended voltage when proteins enter the resolving gel. Gels may also be pre-run (for approximately 1 h) with gel running buffer before loading to improve resolution of proteins.
9. If preparation for Western blotting is not complete before the gel has finished running, the voltage can be reduced to a level that "holds" proteins in the gel but does not allow their further migration.
10. PVDF is easier to handle and has a greater protein binding capacity than nitrocellulose, so it is the membrane we routinely use. PVDF traps proteins on its surface due to hydrophobic interactions and nitrocellulose is hydrophilic. Electro-transfer is less straightforward with nitrocellulose because proteins may pass through this membrane if blotting is performed for too long and nitrocellulose is too fragile to be subjected to repeat probing with antibodies. If proteins are to be visualized with fluorescent stains, PVDF is more suitable because nitrocellulose is explosive and so cannot be placed near a UV light source.
11. No single blocking agent (i.e., BSA or milk) is ideal as the choice of blocker depends on the antigen and the type of detection label used, and each antibody–antigen pair is unique. Inadequate amounts of blocking agent will result in excessive background staining and a reduced signal:noise ratio, and excessive amounts of blocking agent may mask antibody:antigen interactions. We typically use 2.5–5 % (w/v) BSA or milk to block membranes. There are, however, some exceptions to these rules. For phosphotyrosine antibodies milk should not be used as a membrane-blocking agent.
12. Antibodies often require optimization to effectively detect proteins. Antibody concentrations and washing of the membrane after antibody incubations are usually tailored to the protein being examined, as well as the type of sample under investigation. To determine antibody concentrations, incubation time periods, and washing patterns optimal for the detection

of a particular protein, the manufacturer's recommendations should be initially followed. We routinely perform antibody titers against both primary and secondary antibodies to determine concentrations which allow the clear identification of proteins. Generic incubation times of 2 h at room temperature or overnight at 4 °C for primary antibodies and 2 h at room temperature for secondary antibodies are used initially before optimization.

13. We have found several washing patterns to be effective for most proteins. These include three 15-min, seven 5-min, or two 1-h washes.
14. When imaging proteins conjugated to fluorescent tag-labelled antibodies, it is advisable to use PBS for a final wash step before scanning to remove excess detergent that may interfere with the fluorescent signal.
15. A variety of ECL systems are available. In our laboratory, we find that the Pierce ECL kit is reliable as it consistently allows the detection of proteins at a wide range of expression levels. Other ECL systems which allow the detection of proteins of very low abundance (ECL Plus, *SuperSignal West Pico, SuperSignal West Dura, and SuperSignal West Femto kits from Pierce; ECL Plus kit from GE Healthcare*) are also available.
16. X-ray film has a limited dynamic range that does not allow the linear capture of luminescent signals emitted from proteins on membranes. Gel/blot documentation/imaging systems are superior to X-ray film for protein imaging because they are automated to adjust the signal:noise ratio to achieve an optimal signal for fluorescent or luminescent signals from proteins on membranes.
17. Before heating the water bath to 80 °C, overturn a large beaker into the water, ensuring that the base of the beaker is just below the water line. This beaker will serve as a support for the sealed container during the stripping procedure, and is also a safety measure to prevent scalding.

Acknowledgment

This work was funded by the Medical Research Council (MRC).

References

1. Li W, Hayashida Y, Chen YT, Tseng SC (2007) Niche regulation of epithelial stem cells at the limbus. Cell Res 17:26–36
2. Davanger M, Evensen A (1971) Role of the pericorneal papillary structure in renewal of corneal epithelium. Nature 229:560–561
3. Kulkarni BB, Tighe PJ, Mohammed I, Yeung AM, Powe DG, Hopkinson A et al (2010) Comparative transcriptional profiling of the limbal epithelial crypt demonstrates its putative stem cell niche characteristics. BMC Genomics 11:526

4. Majo F, Rochat A, Nicolas M, Jaoudé GA, Barrandon Y (2008) Oligopotent stem cells are distributed throughout the mammalian ocular surface. Nature 456:250–254
5. Schlotzer-Schrehardt U, Kruse FE (2005) Identification and characterisation of limbal stem cells. Exp Eye Res 81:247–264
6. Schermer A, Galvin S, Sun TT (1986) Differentiation-related expression of a major 64 K corneal keratin in vivo and in culture suggests limbal location of corneal epithelial stem cells. J Cell Biol 103:49–62
7. Chen Z, de Paiva CS, Luo L, Kretzer FL, Pflugfelder SC, Li DQ (2004) Characterization of putative stem cell phenotype in human limbal epithelia. Stem Cells 22:355–366
8. Higa K, Shimmura S, Miyashita H, Kato N, Ogawa Y, Kawakita T et al (2009) N-cadherin in the maintenance of human corneal limbal epithelial progenitor cells in vitro. Invest Ophthalmol Vis Sci 50:4640–4645
9. Nakatsu MN, Ding Z, Ng MY, Truong TT, Yu F, Deng SX (2011) Wnt/β-catenin signaling regulates proliferation of human cornea epithelial stem/progenitor cells. Invest Ophthalmol Vis Sci 52:4734–4741
10. Truong TT, Huynh K, Nakatsu MN, Deng SX (2011) SSEA4 is a potential negative marker for the enrichment of human corneal epithelial stem/progenitor cells. Invest Ophthalmol Vis Sci 52:6315–6320
11. Thomas PB, Liu YH, Zhuang FF, Selvam S, Song SW, Smith RE et al (2007) Identification of Notch-1 expression in the limbal basal epithelium. Mol Vis 13:337–344
12. Zhou SY, Zhang C, Baradaran E, Chuck RS (2010) Human corneal basal epithelial cells express an embryonic stem cell marker OCT4. Curr Eye Res 35:978–985
13. Chen B, Mi S, Wright B, Connon CJ (2010) Investigation of K14/K5 as a stem cell marker in the limbal region of the bovine cornea. PLoS One 5:e13192
14. Notara M, Alatza A, Gilfillan J, Harris AR, Levis AR, Schrader S et al (2010) In sickness and in health: corneal epithelial stem cell biology, pathology and therapy. Exp Eye Res 90:188–195
15. Colabelli Gisoldi RA, Pocobelli A, Villani CM, Amato D, Pellegrini G (2010) Evaluation of molecular markers in corneal regeneration by means of autologous cultures of limbal cells and keratoplasty. Cornea 29(7):715–722
16. Shortt AJ, Secker GA, Rajan MS, Meligonis G, Dart JK, Tuft SJ et al (2008) Ex vivo expansion and transplantation of limbal epithelial stem cells. Ophthalmology 115:1989–1997
17. Harkin DG, Barnard Z, Gillies P, Ainscough SL, Apel AJ (2004) Analysis of p63 and cytokeratin expression in a cultivated limbal autograft used in the treatment of limbal stem cell deficiency. Br J Ophthalmol 88:1154–1158
18. Rama P, Bonini S, Lambiase A, Golisano O, Paterna P, De Luca M et al (2001) Autologous fibrin-cultured limbal stem cells permanently restore the corneal surface of patients with total limbal stem cell deficiency. Transplantation 72:1478–1485
19. Kenyon KR, Tseng SC (1989) Limbal autograft transplantation for ocular surface disorders. Ophthalmology 96:709–722
20. Utheim TP, Raeder S, Utheim OA, Cai Y, Roald B, Drolsum L et al (2007) A novel method for preserving cultured limbal epithelial cells. Br J Ophthalmol 91:797–800
21. Kurien BT, Scofield RH (2006) Western blotting. Methods 38:283–293
22. Lowry OH, Rosebrough NJ, Farr AL, Randall RJ (1951) Protein measurement with the Folin phenol reagent. J Biol Chem 193:265–275
23. Hames BD (ed) (1998) Gel electrophoresis of proteins—a practical approach, 3rd edn. Oxford University Press, Oxford, UK
24. Mathews ST, Plaisance EP, Kim T (2009) Imaging systems for westerns: chemiluminescence vs. infrared detection. Methods Mol Biol 536:499–513

Part III

Corneal Cell Culture

Chapter 7

The Culture of Limbal Epithelial Cells

Tor Paaske Utheim, Torstein Lyberg, and Sten Ræder

Abstract

The transplantation of cultured limbal epithelial cells (LEC) has since its first application in 1997 emerged as a promising technique for treating limbal stem cell deficiency. The culture methods hitherto used vary with respect to preparation of the harvested tissue, choice of culture medium, culture time, culture substrates, and supplementary techniques. In this chapter, we describe a procedure for establishing human LEC cultures using a feeder-free explant culture technique with human amniotic membrane (AM) as the culture substrate.

Key words Limbal epithelial cell culture, Amniotic membrane, Limbal explant culture, Ocular surface reconstruction, Limbal stem cell deficiency, Ex vivo expansion of limbal epithelial cells, Corneal regeneration

1 Introduction

Limbal stem cell deficiency (LSCD) may result from a variety of corneal disorders, including chemical and thermal burns, infections, and autoimmune diseases. The symptoms of LSCD include irritation, epiphora, blepharospasms, photophobia, pain, and decreased vision [1]. There are a number of treatment options ranging from nonsurgical treatments to various forms of surgery, including transplantation of amniotic membrane (AM) and different cell types cultured on a range of substrates. Ex vivo expansion of limbal epithelial cells (LEC) involves the culture of LEC harvested either from the patient [2], a living relative [3], or a cadaver [4] on a substrate in the laboratory and the transfer of cultured tissue onto the cornea(s) of patients presenting LSCD.

There are several advantages of this method: only a small biopsy from a healthy eye is needed for reconstruction of the cornea, minimizing the risk of stem cell failure in the donor eye [5]. This allows a further biopsy to be obtained if necessary [6]. Furthermore, compared to procedures that involve direct transplantation of

Bernice Wright and Che J. Connon (eds.), *Corneal Regenerative Medicine: Methods and Protocols*, Methods in Molecular Biology, vol. 1014, DOI 10.1007/978-1-62703-432-6_7,

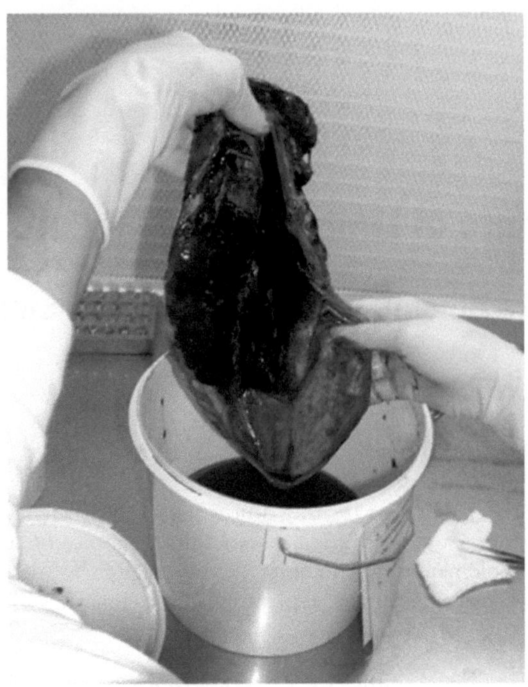

Fig. 1 Preparation of the placenta

limbal tissue, few [7] or no antigen presenting cells [8] are contained in cultivated corneal epithelial sheets, reducing the immune response when allografts are used.

The use of ex vivo cultured LEC in the treatment of LSCD is well characterized, and this procedure has placed ophthalmology at the forefront of regenerative medicine [2–4, 6, 9–55]. The overall success rate of ex vivo expansion of LEC for reconstruction is 74.4 % (661/889 eyes). Culture methods for LEC vary with respect to preparation of the harvested tissue, choice of culture medium, culture time, substrates, and supplementary techniques, including the use of a feeder layer, hypoxia, and air-lifting.

In the present chapter, we describe a protocol for establishing human LEC cultures by a feeder-free explant culture technique using human AM as the culture substrate. Our method is based on a clinically successful protocol first reported by Tsai et al. [56] with minor modifications. The method has since, in more or less modified forms, been applied at several centers around the world and has produced promising clinical results. Various aspects, including strengths and weaknesses, of the protocol are discussed in Subheading 4.

2 Materials

2.1 Amniotic Membrane Preparation and Preservation

1. Chlorhexidine (Fresenius Kabi AG, Bad Homburg, Germany).
2. 70 % (v/v) Ethanol.
3. Phosphate-buffered saline (PBS).
4. Antibiotics: 50 U/mL penicillin (Sigma-Aldrich, St. Louis, MO, USA), 50 μg/mL streptomycin (Sigma), and 2.5 μg/mL amphotericin B (Sigma).
5. Dulbecco's modified Eagle's medium (DMEM) (Sigma).
6. Glycerol.
7. Class II biosafety cabinet (HERAsafe, Thermo Scientific, Waltham, MA, USA).
8. Radiation-sterilized nitrocellulose membrane (VWR International, Radnor, PA, USA).
9. 3 L Plastic bottle for collection of fetal membranes.
10. Cryopreservation vials for preserving AM (Sarstedt, Nümbrecht, Germany).
11. Sterilized scissors.
12. Moorfield forceps or round-ended forceps.
13. Freezer (−80 °C).

2.2 Human Limbal Epithelial Cell Culture

1. PBS.
2. DMEM.
3. 4-(2-Hydroxyethyl)-1-piperazineethanesulfonic acid (HEPES)-buffered DMEM containing sodium bicarbonate and Ham's F12 (1:1).
4. Medium supplements: 2 ng/mL human epidermal growth factor (hEGF), 0.5 % (v/v) dimethyl sulfoxide (DMSO), 5 μg/mL insulin, 5 μg/mL transferrin, 5 ng/mL selenium, 3 ng/mL hydrocortisone, 30 ng/mL cholera toxin (Enzo Life Sciences, Farmingdale, NY, USA), 50 μg/mL gentamycin, 1.25 μg/mL amphotericin B, and 50 U/mL penicillin.
5. Dispase II (1.2 U/mL).
6. Magnesium- and calcium-free Hank's balanced salt solution (HBSS).
7. 100 % FBS.
8. Class II biosafety cabinet.
9. CO_2 incubator (*see* **Note 1**) (Heraeus HERAcell® 240, Thermo Scientific).
10. Phase contrast microscope (Leica Microsystems, Wetzlar, Germany).

11. Pipettes.
12. Vials to contain culture medium.
13. Plastic dishes to prepare explants.
14. Surgery kit containing scissors, needle holder, 6–0 (thickness) monofilament sutures, straight steel blade, beveled-edged steel blade, sterile surgical blades, Moorfield forceps, tying forceps, Beaver sclerotome, and surgical sponges.
15. 6-well culture plates with 24 mm diameter culture plate inserts (Netwell, 74 μm mesh size polyester membrane, Costar, Corning Inc., Acton, MA, USA).

2.3 Storage of LEC Cultured on Amniotic Membrane

1. Minimum essential medium (MEM) (Life Technologies, Carlsbad, CA, USA) with 12.5 mM HEPES.
2. 7.5 % (w/v) Sodium bicarbonate.
3. Antibiotics: 50 μg/mL gentamycin, 100 μg/mL vancomycin (Abbott, North Chicago, IL, USA), and 2.5 μg/mL amphotericin B.
4. Disposable forceps (Unomedical, McAllen, TX, USA).
5. 50 mL Glass bottles (OneMed) (*see* **Note 2**).

3 Methods

Carry out all procedures at room temperature unless otherwise specified (*see* **Note 3**).

3.1 Preparation of Human Amniotic Membrane

1. Ensure written, informed consent is obtained before placental tissue is donated.
2. Perform serologic tests to ensure absence of human immunodeficiency virus (HIV), hepatitis virus type B and C, and syphilis (*see* **Note 4**).
3. Wash a 3 L plastic bottle with chlorhexidine and spray the inside of the bottle and the lid with 70 % (v/v) ethanol. Dispense 500 mL PBS supplemented with 50 U/mL penicillin, 50 μg/mL streptomycin, and 2.5 μg/mL amphotericin B into the bottle. Collect the placenta with attached fetal membranes (which includes AM) immediately after elective cesarean delivery, and place the tissue into the sterilized bottle containing antibiotics (*see* **Note 5**).
4. Store the placenta and fetal membranes at 4 °C until further processing (Fig. 1) (*see* **Note 6**).
5. Eliminate fetal membranes from the placenta using a pair of scissors (*see* **Note 7**).

Fig. 2 The chorion is removed, leaving the transparent amniotic membrane attached to the nitrocellulose paper

6. Wash fetal membranes with washing solution (PBS supplemented with 50 U/mL penicillin, 50 μg/mL streptomycin, and 2.5 μg/mL amphotericin B) until all residual blood clots are washed off completely.
7. Mix thoroughly equal volumes of DMEM and glycerol in a class II biosafety cabinet, and prepare aliquots of the solution in sterile vials for cryopreservation (*see* **Note 8**).
8. Spread fetal membranes uniformly on sterilized 0.45 μm nitrocellulose membranes with the vascularized chorion facing up. Using blunt forceps, remove the chorion, leaving the AM attached to the nitrocellulose membrane (Fig. 2) (*see* **Notes 9** and **10**).
9. Keep the AM wet by intermittent addition of washing solution.
10. Cut the AM into pieces with dimensions of ~2.5 × 2.5 cm (*see* **Note 11**).
11. Roll the pieces of nitrocellulose membrane with AM attached using Moorfield forceps. Insert the AM/nitrocellulose membrane rolls into sterile vials containing sufficient cryopreservation medium to cover the entire membrane roll.

Fig. 3 The amniotic membrane is coupled to the insert membrane of a Costar Netwell culture plate insert (24 mm diameter, polyester, 74 μm pore size) using 6-0 monofilament suture

12. Repeat **steps 9–11** until the AM is completely sectioned (*see* **Note 12**).
13. Label the vials with donor information (donors are anonymous), the date of preparation, and a lot number (*see* **Note 13**).
14. Store vials at −80 °C (*see* **Note 14**).

3.2 Establishment of Human Limbal Epithelial Cell Cultures

1. Thaw the AM at room temperature (*see* **Note 15**).
2. Wash the AM three times in PBS to eliminate residual glycerol. Transfer the AM with the epithelial side facing up (*see* **Note 10**) to the polyester membrane of the 6-well culture plate insert using tying forceps, and flatten the AM.
3. Fasten the AM to the polyester membrane of the culture plate insert using monofilament sutures, and add ~1.5 mL DMEM to the well to keep the AM moist (Fig. 3) (*see* **Note 16**).
4. Obtain corneoscleral tissue from an Eye Bank (*see* **Note 17**).
5. Rinse corneoscleral rims three times with DMEM containing 50 μg/mL gentamycin and 1.25 μg/mL amphotericin B.
6. Carefully remove excessive conjunctiva, Tenon's capsule, sclera, iris, and corneal endothelium using a straight steel blade and a Beaver sclerotome.
7. Place the remaining corneolimbal tissue in a culture dish.

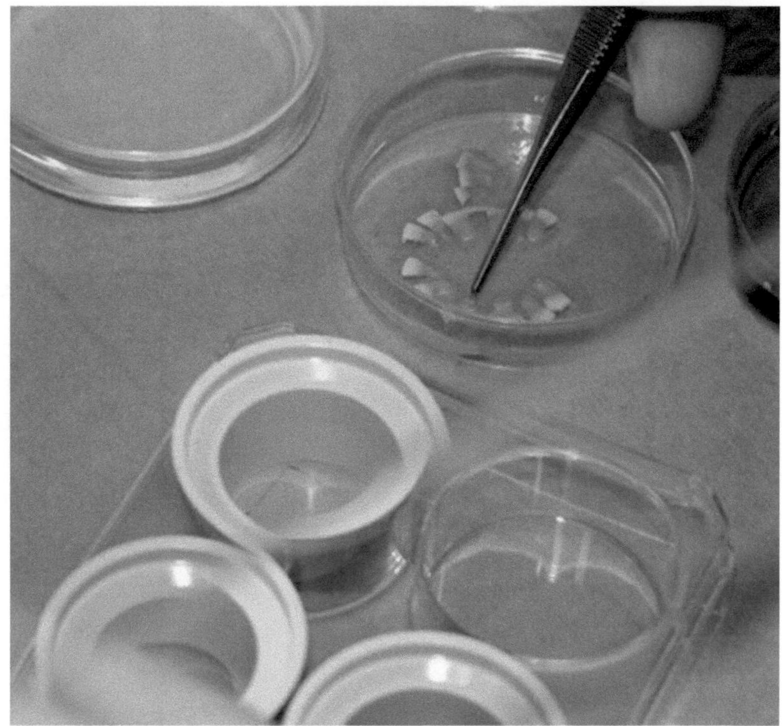

Fig. 4 The explant culture method. The limbal ring is divided into limbal explants and cultured on the amniotic membrane

8. Add Dispase II (1.2 U/mL) in magnesium- and calcium-free HBSS and incubate the limbal tissue for 10 min in a humidified incubator maintained at 37 °C with a 5 % CO_2 supply.
9. Rinse limbal tissue with DMEM containing 10 % (v/v) FBS to stop the enzymatic reaction.
10. Using a beveled-edged steel blade, cut corneoscleral ring explants (*see* **Note 18**) of ~2 × 2 mm (Fig. 4) (*see* **Note 19**).
11. Place limbal explants onto the center of each AM insert with the epithelial side facing down [40] (Fig. 5) (*see* **Note 20**).
12. Add 100 % FBS drop wise over the explant and incubate for at least 2 h (*see* **Note 21**).
13. Carefully add (*see* **Note 22**) 1–2 mL HEPES-buffered DMEM containing sodium bicarbonate and Ham's F12 (1:1) supplemented with 5 % (v/v) FBS, 0.5 % (v/v) DMSO, 2 ng/mL hEGF, 5 μg/mL insulin, 5 μg/mL transferrin, 5 ng/mL selenium, 3 ng/mL hydrocortisone, 30 ng/mL cholera toxin, 50 μg/mL gentamycin, and 1.25 μg/mL amphotericin B (*see* **Note 23**) to submerge explants (*see* **Note 24**).
14. Place the 6-well plate containing explants into the incubator (*see* **Note 25**).

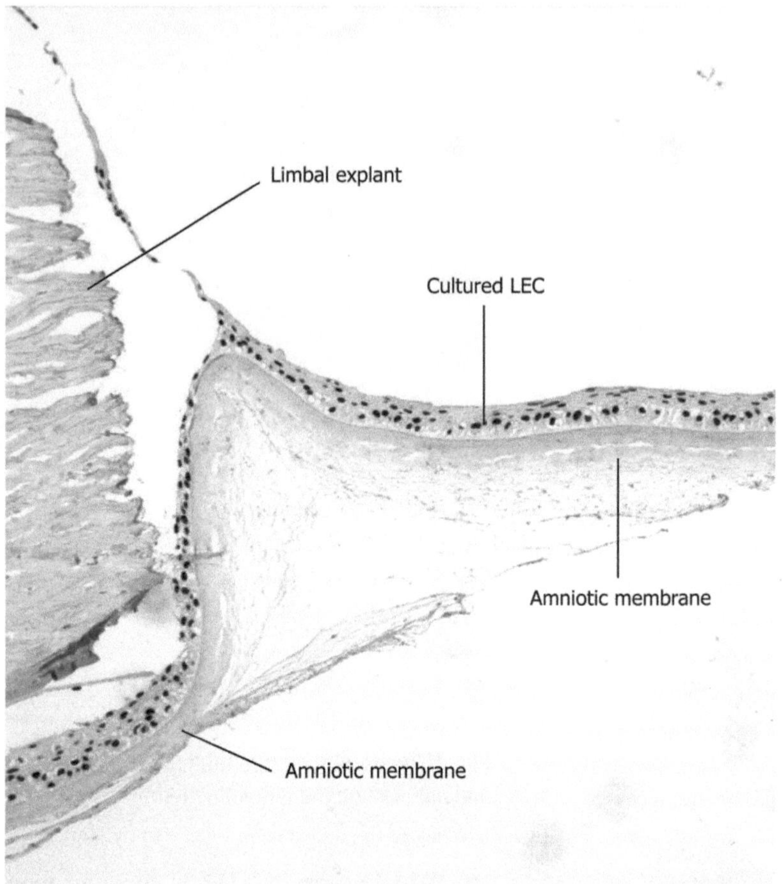

Fig. 5 Limbal epithelial cells growing from the limbal explant with the epithelium facing downwards onto the amniotic membrane

15. After 2 days, add 2–3 mL of medium (*see* Subheading 3.2, **step 13**) carefully along the inner wall of the culture well (*see* **Note 26**).
16. Change the medium (*see* Subheading 3.2, **step 13**) every 2–3 days during the culture period (2–3 weeks) (*see* **Note 27**).

3.3 Storage of Cultured Human Limbal Cells

1. Dispense 50 mL storage medium containing HEPES-buffered MEM supplemented with 50 μg/mL gentamycin, 100 μg/mL vancomycin, and 2.5 μg/mL amphotericin B into sterilized 50 mL glass containers (*see* **Note 28**).
2. Release the polyester mesh membrane with cultured epithelium attached using a straight steel blade. Tie the edge of the polyester membrane and the rubber cap using a 6-0 (thickness) monofilament suture and suspend the tissue in glass bottles containing storage medium.

3. Seal glass bottles using a rubber cap.
4. Store cultured LEC at 23 °C as required; although these cell sheets should not be preserved for longer than 1 week before transplantation onto the cornea (*see* **Note 29**).

4 Notes

1. A copper incubator is preferable as the copper releases copper oxide, which destroys microbes present in the chamber.
2. Alternatively, a plastic container with a septum for aspiration of medium to enable microbial testing may be used.
3. Make certain that the procedures are in agreement with the local ethical committee and regulatory legislation.
4. Serological screening is generally a part of routine presurgical testing at the Maternity Department of the hospital and hence completed at the time of tissue collection. Testing for human T-lymphotropic virus type I and type II (HTLV-I and HTLV-II) should be considered [49, 57], although it is not commonly performed [39, 58–60]. To be certain to cover the "window period" of infections, the donor is tested serologically 6 months after the first blood test [61]. Ensure that written information given to the donor prior to the AM donation states the need for two blood samples approximately 6 months apart to test for microbiological agents (e.g., hepatitis A and B, syphilis, HIV, HTLV-I, and HTLV-II).
5. Placental tissue should be harvested after elective cesarean section rather than during vaginal labor to minimize the risk of contamination and structural defects associated with stretching of the membrane during labor and delivery. Placenta collected after natural vaginal delivery may be contaminated by normal vaginal flora and a number of pathogens, including herpes virus and chlamydia [62].
6. Fetal membranes should be processed within 24 h of tissue collection.
7. Ideally, media and washing solutions needed for the preparation of AM should be produced within 10 days prior to use [60]. Sterility checks and pH measurements of media before use are recommended by some investigators [60].
8. Amniotic membrane cryopreservation solution should be freshly prepared and transferred to sterile plastic vials [59, 60]. A 1:1 mixture glycerol: DMEM is the most frequently used cryopreservation medium for AM [63]. Glycerol has antibacterial properties [64]. Shortt et al. [65] found a higher percentage of stem cell markers expressed in cultured LEC on AM

which had been cryopreserved with only HBSS compared to the standard protocol with equal amounts of glycerol and DMEM. Moreover, the additional use of peracetic acid (PAA) or antibiotics to decontaminate the AM did not appear to have any effect on cell density, cell confluence, or the expression of stem cell markers [65]. The use of PAA is a useful alternative to antibiotics as it destroys all contaminants, including viruses, as opposed to antibiotics which target only bacterial and fungal pathogens [65].

9. Using the described method, the side facing the chorion is the stromal aspect of the AM, and the side of the fetal membrane attached to the nitrocellulose membrane is the AM epithelium. An alternative approach for preparing AM samples is to separate the amnion and the chorion before the AM is attached to nitrocellulose membrane. Any leftover chorion attached to the AM and blood clots are gently peeled off the epithelial cell layer using round-ended forceps. With the epithelial side surface up, the AM is uniformly spread on sterilized 0.45 μm nitrocellulose membranes, allowing the tissue to adhere. Care is taken to avoid folds, trapping air bubbles, or acquiring tears in the AM.
10. To distinguish the sides of the AM, use surgical sponges. Unlike the epithelial side, the stromal side of the AM is sticky.
11. The size of the AM pieces depends on the culture system used. In the present protocol, AM is intact, i.e., containing the epithelial layer. There is currently a debate over whether AM should be de-epithelialized/denuded (partial or complete removal of the single layer of columnar epithelial cells on AM by either enzymatic digestion, chemical treatment, or physical scraping) prior to LEC culture, or if this substrate should remain intact. Several studies have shown that LEC cultured on intact AM develop a more stem cell-like phenotype compared with LEC cultured on denuded AM [66–68]. Koizumi and co-workers [69] showed that LEC cultured on denuded AM formed a more stratified and differentiated epithelium and exhibited a higher number of desmosomes and hemidesmosomes than when cultured on intact AM. Only seven clinical studies (sub-studies excluded) have applied intact AM [3, 16, 32, 33, 39, 45, 47] as a culture substrate, whereas 25 clinical studies used denuded AM to culture LEC [6, 12, 17–21, 26–28, 34–38, 42–44, 49, 51–55, 58].
12. Provided that the harvested fetal membrane is intact, typically 70–80 AM samples measuring ~2.5×2.5 cm can be prepared from a single placenta.
13. For future documentation, check the AM batch by performing microbiological analysis of the medium in a sample vial.

14. The recommended storage period for AM varies from 3 months [70], 6 months [59], 12 months [71], 24 months [72] to an indefinite period of time [71]. Thomasen et al. [72] compared different storage periods of AM (4, 15, and 24 months). They did not find any significant impairment of sterility, histology, or biological properties of AM during storage. Reported storage temperature for AM is almost exclusively −80 °C, but −70 °C has also been used [26, 36].

15. LEC maintain their in vivo properties of slow cycling/label retention [73–75] and undifferentiated phenotype when cultured on AM [73, 76–78]. Cross-linking with carbodiimide (*see* Chapter 10) is a novel technology to enhance the mechanical and thermal stability, optical transparency, and resistance to collagenase digestion of AM [79]. Freeze-dried [80] and air-dried AM [81] have recently become commercially available and have the advantage of being ready-to-use products [82]. By comparing air-dried AM with cryopreserved AM, Thomasen and colleagues [82] found air-dried AM to be inferior as cultured limbal explants displayed a lower outgrowth rate, decreased release of soluble wound-healing modulating factors, and inferior preservation of basement membrane components.

    There is currently much research performed to identify alternative culture substrates, as AM presents a number of limitations: (a) it is semiopaque, (b) it is time-consuming to prepare, (c) there is a risk of transfer of pathogens from AM, and (d) AM presents inter-donor and intra-donor variations [83–85], which influence the outcome of its clinical use.

    There are several alternatives to AM for the culture of LEC, including fibrin [86], silk fibroin (*see* Chapter 11) [87, 88], silk fibroin mixed with polyethylene glycol [87], plastic [29], laminin-coated compressed collagen gel (*see* Chapter 9) [89], plastic compressed collagen [90, 91], human keratoplasty lenticules [92], nanofiber scaffolds (*see* Chapters 12 and 13) [93], poly(lactide-co-glycolide) electrospun scaffolds (*see* Chapter 12) [94], poly-ε-caprolactone electrospun scaffolds [95], chemically cross-linked hyaluronic acid-based hydrogels [96], contact lenses [2, 3, 30, 97], culture inserts [29], collagen IV-coated plates [98], collagen membranes [99, 100], chitosan matrix [101], chitosan silver matrix [101], chitosan gold matrix [101], temperature responsive culture dishes [102–104], Mebiol Gel (thermo-reversible polymer gel) [105, 106], Matrigel (reconstituted basement membrane extract) [100, 107, 108], petrolatum gauze [2], and recombinant human cross-linked collagen scaffold [109]. Of all these substrates, only fibrin [13, 19, 31, 40, 46, 48], contact lenses [2, 30], culture inserts [29], and petrolatum gauze [2] have been used clinically.

16. We have found the use of the polyester membrane of Netwell culture plate inserts [110] to be superior to mixed cellulose ester membrane [111] as the polyester membrane is able to: (a) withstand the tension of sutures, (b) keep the AM distended, and (c) easily release the AM attached to the culture plate insert prior to surgery. In contrast, the cellulose ester membrane was shown to be fragile and tended to rupture during the process of fastening the AM [111].

    Amniotic membranes may also be attached with the epithelial surface facing upwards, with a circumferential suture to the bottom of a cell culture insert from which the base has been removed [112]. In our experience, however, it is crucial that AM has a firm support to obtain optimal growth of LEC. Moreover, as AM retracts during culture, it is important to fasten this tissue with a running suture with at least six anchoring points along the circumference. We, therefore, in 2008, developed culture systems to ensure firm support, no retraction during culture, and easy handling of AM (Figs. 6 and 7).

17. Corneolimbal tissue might be harvested from a patient with LSCD, especially in cases of strict unilateral disease. If both eyes are affected or the underlying condition is a systemic illness, such as Stevens–Johnson syndrome, corneolimbal tissue donated either from a living relative or a deceased person should be considered [61]. If fresh tissue is used, the excised biopsy will typically be ~2 × 2 mm in size (including ~1 mm of clear cornea) and be ~150–200 μm in thickness. The conjunctiva should be excised just behind the palisades of Vogt [59].

    Fresh limbal tissue can be transported at ambient temperature to the laboratory in sterile 1.5 mL microfuge tubes containing 1 mL culture medium [59]. It is recommended that the limbal biopsy is processed the same day the tissue is harvested. The tissue can be stored for up to 24 h at 4 °C [54] or at ambient temperature (if necessary during transit) without much reduction in cell viability [59]. In some clinical studies, all tissue donors, whether autologous or allogeneic, were reported to be screened for hepatitis B and C, syphilis, HIV-1, HIV-2, and HTLV prior to use [48, 49].

    There are several factors which might determine the success of LEC cultures, including donor age [113–116], donor gender [114, 117], death to enucleation time [117, 118], and death to culture time [119]. Information about these parameters should therefore be provided. Corneas can be stored in HEPES-buffered MEM supplemented with antibiotics at ambient temperature prior to and during transportation to the laboratory. Four studies have investigated the effect of harvesting sites on the culture of LEC and concluded that the superior region of the eye [114, 120–122] is suitable for

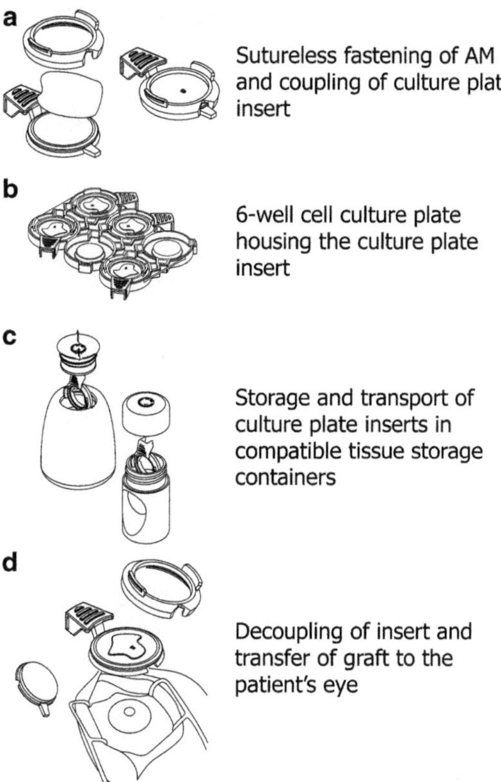

Fig. 6 Ræder and Utheim's culture, storage/transportation, and surgery devices for treating limbal stem cell deficiency. Courtesy of Ingunn Hesselberg, Håkon Raanes, and Erik Falk-Petersen, The Oslo School of Architecture and Design, Norway

harvesting limbal tissue. The highest colony forming efficiency (*see* Chapter 3) was generated by biopsies from the superior and inferior region [114, 121], whereas the superior region yielded the highest epithelial cell thickness of cultured limbal explants [120].

18. The explant method is most commonly used in clinical studies applying cultured LEC [4, 6, 12, 16, 17, 20, 26, 28, 30, 32–34, 37–39, 41, 42, 44, 45, 47, 53, 54, 58]. An alternative technique is cell suspension [2, 3, 13, 18, 19, 22, 27, 29, 31, 38, 40, 43, 44, 46, 48, 49, 54], in which single LEC are released from the limbal tissue after enzyme treatment. Digestion of limbal tissue with dispase [19], which cleaves the basement membrane [123], fails to isolate all limbal basal epithelial progenitors while additional digestion with trypsin/ethylenediaminetetraacetic acid (EDTA) [49] disrupts the close interaction between limbal basal progenitors and their niche cells [124, 125].

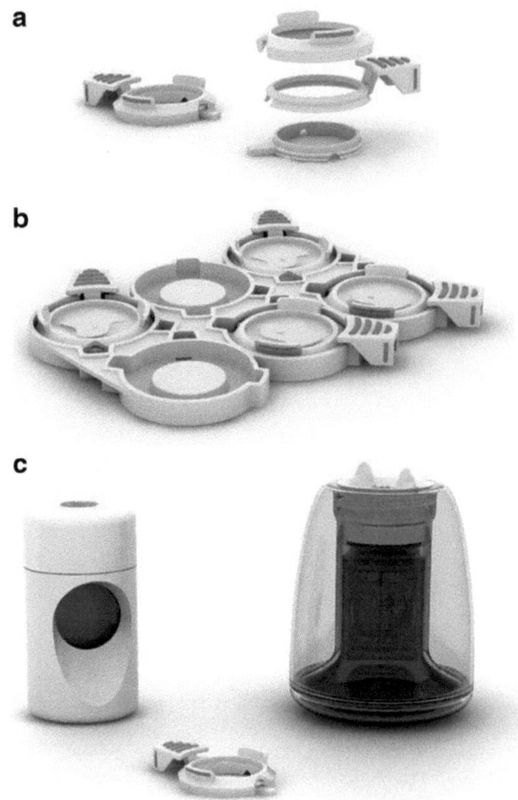

Fig. 7 A prototype for culture and storage of limbal epithelial cells developed by Utheim, Raeder, and co-workers in 2008. (**a**) Culture plate insert specifically developed for quick, sutureless fastening of the amniotic membrane. (**b**) 6-well cell culture plate housing the culture plate insert. (**c**) Culture plate insert and compatible tissue storage containers. Courtesy of Ingunn Hesselberg, Håkon Raanes, and Erik Falk-Petersen, The Oslo School of Architecture and Design, Norway

Interestingly, Cristovam et al. [126] demonstrated that despite the use of one third of the entire limbal circumference to generate a cell suspension, neither holoclones nor meroclones were observed. In contrast, cell suspensions that were co-cultured with a 3T3 feeder layer formed such clones. Chen et al. [124] recently confirmed the crucial role of 3T3 cells if limbal tissue is to be dissociated into single cells. In defined keratinocyte serum-free medium, single cells derived from dispase or collagenase followed by trypsin/EDTA treatment did not generate any clonal growth, unless 3T3 cells were added. In contrast, cells released from the remaining stroma by collagenase without using trypsin/EDTA resulted in great clonal growth [124]. As 3T3 cells are animal-derived, their inclusion in the culture protocol has raised considerable concern(*see* Chapter 1) [127].

Seven recent clinical reports that involved cultured LEC have demonstrated promising results based on a culture protocol devoid of animal components (e.g., 3T3 feeder layer) [6, 14, 20, 30, 38, 44, 47]. Noteworthy, all these protocols used explants instead of cell suspension. As no highly successful animal free culture protocol applying cell suspension is described, there is a strong rationale for using explants, which provide favorable conditions for stem/progenitor replication [128]. As the use of limbal explants does not involve enzymatic treatment, the loss of residual stem cells during processing is avoided [59]. Mariappan et al. [59] stated that the use of limbal explants to treat LSCD was the best option if tissue has to be harvested from a healthy area of a partially damaged eye [41].

Stem cells reside in a specialized microenvironment known as a niche [129]. There is increasing evidence that neighboring cells, intercellular interactions, and other local environmental factors, such as extracellular matrix and signalling molecules, control stem cell function [130]. The use of explants may permit maintenance of the LSC niche. The proliferative status of cells remaining in the explant has been shown, however, to abate when epithelial sheet outgrowth from the explant reached confluence [131]. Moreover, Kolli et al. [132] revealed a steady decline in a wide range of stem cell properties in the outgrowth of limbal explants with increasing distance from the limbus. These findings support the notion that the proximity of stem cells to their niche environment is essential in maintaining their undifferentiated state.

This observation is supported by Zhang et al. [133] who demonstrated that ΔNp63α-positive cells were mainly centered around explants, whereas peripheral cells were ΔNp63α-negative. Furthermore, Tan and co-workers [134] described a gradual decline in outgrowth potential of explants as a consequence of intrastromal invasion by LEC in explant cultures. Such intrastromal invasion is suggested to result in epithelial–mesenchymal transition [135]. During ex vivo expansion on intact AM, Li et al. [135] showed that the clonogenicity of epithelial cells from the explant declined from passage 1 (P1) to P3, eventually resulting in no colony formation at P3. These findings suggested that epithelial progenitor cells are potentially exhausted during the continuous migration from the limbal explant over a 6-week period.

Several studies have compared the explant and cell suspension methods:

(a) Shimazaki et al. [44] concluded that both cultivation strategies produced epithelial sheets with a corneal phenotype that were suitable for transplantation. Corneas treated with the explant method, however, tended to suffer more severe postoperative complications.

(b) Kim et al. [119] showed that in cell suspension culture, cells were smaller, more compact, and more uniform compared to cells from an explant culture. No difference in phenotype or bromodeoxyuridine-label retention was revealed between the two experimental groups.

(c) Koizumi et al. [136] concluded that both cell suspension and explant culture methods produced a healthy LEC layer. The cell suspension culture, however, presented significantly more desmosomal junctions and smaller intercellular spaces than the explant cultured cells.

There are several arguments for the use of cell suspension, including:

(a) The culture time might be shorter [137] compared with explant methods, which require 2–3 days to achieve visible growth [132].

(b) Zhang et al. [133] showed that the cell suspension technique was superior to the explant culture technique in terms of producing an undifferentiated epithelium (judged by the content of ΔNp63α labeling).

19. The size of limbal explants used for culture varies in the literature from 0.4 to 0.5 mm [59] to 3×3 mm [37, 38]. In some cases the explant is cut into small pieces (shredded explants) prior to culture [6, 12, 20, 26, 32, 41, 42, 44, 51–53, 55, 58]. Our research group has compared the use of 1×1 mm limbal explants with 3×1 mm limbal explants and concluded that both explant sizes produced similar cultured sheets [138]. Therefore, a corneoscleral ring can generate approximately 36 cultures rather than 12. In a clinical setting, the use of a small biopsy minimizes the risk of inducing LSCD in the donor eye [5]. The option of small biopsies reduces the threshold for harvesting more than one biopsy to increase the likelihood of successful cultures [49] and increase the clinical outcome [6, 48].

20. Placing limbal explants with the epithelial side facing down ensures the highest possible expression of the putative progenitor/stem cell markers p63 and ΔNp63α [111]. Rama et al. [48] found that cultures in which p63-bright cells constituted more than 3 % of the total number of clonogenic cells were associated with successful transplantation in 78 % (87/112 eyes) of patients. In contrast, cultures in which such cells made up 3 % or less of the total number of cells were associated with successful transplantation in only 11 % of patients [48].

21. The content of fibronectin and vitronectin in FBS is important for initial attachment of limbal cells [139]. For clinical use, 100 % autologous serum (AS) should replace FBS throughout the protocol to avoid animal-derived products. After informed consent, AS can be prepared by draining venous blood into

10 mL vacutainer tubes without anticoagulants. Sharp centrifugation (3,000×g for 15 min), long clotting time (at least 120 min at ambient temperature), and dilution with Balanced Salt Solution have been recommended to improve the ability of serum to support proliferation, migration, and differentiation of corneal epithelial cells [140].

Interestingly, Liu et al. (2005) demonstrated higher levels of EGF, HGF, and TGF-β after 2 h clotting time compared to 15 min [140]. A g-force as high as 3,000 is advantageous as it: (a) yields a large volume of serum from a full blood sample without inducing hemolysis [141] and (b) reduces membranous platelet remnants in the supernatant, which might induce apoptosis of LEC [142, 143]. The serum is collected and passed through 0.22 μm pore size filters. Aliquots of serum can be stored frozen for up to 3 months [143]. To minimize the risk of transmission of microorganisms, it is strongly recommended that the donor is tested serologically [144]. To reduce the risk of bacterial contamination, no blood should be taken from patients with suspected septicemia [144].

22. Medium should be carefully added with the pipette tip along the inner wall of the culture plate.

23. Cholera toxin, despite its name, has been assessed as a low risk culture component given the degree of purification and dose, but disease transmission remains theoretically possible, albeit unlikely [127]. Isoproterenol mimics the actions of cholera toxin by increasing the level of cyclic adenosine monophosphate in the cytoplasm and might be a possible alternative [145].

    In an attempt to reduce/avoid the use of xenobiotic components in a clinical setting, AS [6, 19, 20, 30, 34, 37–39, 44, 47, 58] has increasingly replaced FBS [2–4, 12, 13, 16, 18, 21, 26–29, 31–36, 38, 40–43, 45, 46, 48, 49, 52–55]. Nakamura et al. [146] found the corneal epithelial sheets cultured in AS- and FBS-supplemented media to be morphologically and phenotypically similar. Interestingly, using a similar protocol as the one described in this chapter, Shahdadfar et al. [147] suggested that DMEM/F12, 10 % human serum, and antibiotics may serve as an equivalent replacement for the commonly used complex medium, hence eliminating the need for hEGF, DMSO, insulin, transferrin, selenium, hydrocortisone, cholera toxin, and FBS.

24. Do not add too much medium on the first day of culture as this increases the likelihood of explant detachment. Too little medium, on the other hand, can dry out the cultures and impede outgrowth of epithelial cells. If the explant detaches, it can sometimes be reattached by repeating the procedure. Without attachment, no outgrowth can be expected. To ensure attachment, Sharma et al. [21] allowed the harvested limbal tissue to

adhere on AM for 10 min, after which a few drops of growth medium were added before incubation for 6–7 h. Thereafter, 2 mL of growth medium were added into the culture well. When cadaveric tissue is used, the limbal explant thickness might vary from 100 μm [148] to full thickness (~700 μm) [111]. Recently, Ghoubay-Benallaoua et al. [149] found superficial limbal explants to be superior to full-thickness explants.

25. Take care to transfer the 6-well plate from the biosafety bench to the incubator in the gentlest way to avoid detachment of the explant. This is especially important during the first days of culture. LEC are routinely cultured in vitro in atmospheric oxygen (~21 %), despite the fact that physiological oxygen levels range from 1 to 13 % [150]. On leaving the stem cell niche, cells are exposed to a higher oxygen concentration, which trigger differentiation [150, 151]. As compared to normoxia, hypoxic culture conditions have been shown to prevent abnormal (epidermal-like) differentiation [152] and increase proliferation [152–154], while preventing differentiation [153, 154] of cultured LEC. Recently, O'Callagan and associates [155] found both 2 and 21 % oxygen to be suboptimal for culture of LEC, whilst 14 % oxygen promoted expansion of rabbit LEC with stem cell characteristics. Differences in culture protocols might help explain the contrasting results.

26. The issue of whether limbal explants should be submerged during the entire culture period or not (air-lifted) is controversial. Air-lifting implies that the level of culture medium is lowered to the level of the surface of the epithelium. Air-lifting is normally [156] performed at the end of the culture period, often during the final week. The duration of air-lifting might vary from 1–2 days [38, 58] to up to 2 weeks [3, 4, 35, 43].

    Arguments for air-lifting include the promotion of: (a) barrier function [148], (b) migration [157], (c) proliferation [157], (d) epithelial stratification [157], and (e) reduction of intercellular spaces [148]. The effects of air-lifting, however, depend on whether AM is intact or denuded [158]. Arguments against air-lifting include induction of squamous metaplasia [136, 159, 161], terminal differentiation [114], and gradual loss of stem cells [114]. In the present protocol, a submerged culture technique is used, which has been applied in approximately two thirds of clinical reports of cultured LEC [2, 3, 6, 13, 16, 18, 20–23, 25, 26, 28–33, 39, 41, 42, 44–48, 51–53, 55]. Air-lifted cultures were identified in 14 clinical studies [4, 19, 27, 34–38, 40, 43, 44, 49, 54, 58].

27. At the time of medium shift, the cultured cells can be observed by phase contrast microscopy. In most clinical studies, the culture time for LEC ranged between 14 and 21 days [2, 6, 12, 13, 16, 18–23, 26–28, 32, 33, 38–40, 42, 44–50, 52, 54, 55,

58]. In six clinical studies the culture time was shorter [12, 26, 29, 30, 53, 55], whereas it was longer in seven [3, 4, 35–38, 43] reports. The longer culture time was associated with the use of the air-lifting technique. Recently, Hayashi et al. [104] demonstrated that culture periods of 28 days were inferior to 15 days with a decrease in both viability and the number of cell layers. LEC were cultured under Good Manufacturing Practice conditions prior to transplantation in four studies [46–49].

28. The conversion to a serum-free medium has several advantages, including: (a) avoiding the risk of transmitting infectious agents such as prions [160], (b) improved reproducibility of results as variation between batches of serum is avoided, and (c) easier downstream processing (i.e., no interference with serum) of products from cultured cells with no risk of degradation of sensitive proteins by serum proteases.

    For storage of cultured LEC, organ culture medium containing DMEM with 7.5 % (w/v) sodium bicarbonate, 8 % (v/v) FBS, 50 μg/mL gentamycin, 100 μg/mL vancomycin, and 2.5 μg/mL amphotericin B can also be used at 23 °C [110, 161, 162]. Interestingly, the conventional storage temperature of organ culture medium (31 °C) and Optisol-GS (4–5 °C) are not appropriate for cultured epithelial cells [163]. Although cultured LEC stored in Optisol-GS at 23 °C resulted in significantly improved morphology and viability compared to storage at 5 °C [164, 165], epithelial sheets stored under these conditions have been shown to remain somewhat inferior compared to those stored in HEPES-MEM at 23 °C [166].

29. Our original storage system can withstand mechanical insults during transportation simulations [167]. Alternative systems with improved features for the storage of cells have been developed (Figs. 6 and 7).

Acknowledgments

The authors thank Astrid Østerud at the Department of Ophthalmology, Oslo University Hospital, Norway; Ingrid Riphagen at the Unit for Applied Clinical Research, Norwegian University of Science and Technology, Norway; Øygunn Aass Utheim and Jon Roger Eidet at the Department of Medical Biochemistry, Oslo University Hospital, Norway; Borghild Roald at the Department of Pathology, Oslo University Hospital, Norway; Edward Messelt at the Institute of Oral Biology, Dental Faculty, University of Oslo, Norway for excellent help and support.

References

1. Dua HS, Azuara-Blanco A (2000) Limbal stem cells of the corneal epithelium. Surv Ophthalmol 44:415–425
2. Pellegrini G, Traverso CE, Franzi AT, Zingirian M, Cancedda R, De Luca M (1997) Long-term restoration of damaged corneal surfaces with autologous cultivated corneal epithelium. Lancet 349:990–993
3. Schwab IR (1999) Cultured corneal epithelia for ocular surface disease. Trans Am Ophthalmol Soc 97:891–986
4. Koizumi N, Inatomi T, Suzuki T, Sotozono C, Kinoshita S (2001) Cultivated corneal epithelial stem cell transplantation in ocular surface disorders. Ophthalmology 108:1569–1574
5. Jenkins C, Tuft S, Liu C, Buckley R (1993) Limbal transplantation in the management of chronic contact-lens-associated epitheliopathy. Eye 7:629–633
6. Basu S, Ali H, Sangwan VS (2012) Clinical outcomes of repeat autologous cultivated limbal epithelial transplantation for ocular surface burns. Am J Ophthalmol 153:643–650, e2
7. Kinoshita S, Koizumi N, Sotozono C, Yamada J, Nakamura T, Inatomi T (2004) Concept and clinical application of cultivated epithelial transplantation for ocular surface disorders. Ocul Surf 2:21–33
8. Lim P, Fuchsluger TA, Jurkunas UV (2009) Limbal stem cell deficiency and corneal neovascularization. Semin Ophthalmol 24: 139–148
9. Sharpe JR, Daya SM, Dimitriadi M, Martin R, James SE (2007) Survival of cultured allogeneic limbal epithelial cells following corneal repair. Tissue Eng 13:123–132
10. Pauklin M, Steuhl KP, Meller D (2009) Characterization of the corneal surface in limbal stem cell deficiency and after transplantation of cultivated limbal epithelium. Ophthalmology 116:1048–1056
11. Pauklin M, Kakkassery V, Steuhl KP, Meller D (2009) Expression of membrane-associated mucins in limbal stem cell deficiency and after transplantation of cultivated limbal epithelium. Curr Eye Res 34:221–230
12. Sahu SK, Das S, Sachdeva V, Sangwan VS (2009) Alcaligenes xylosoxidans keratitis after autologous cultivated limbal epithelium transplant. Can J Ophthalmol 44:336–337
13. Marchini G, Pedrotti E, Pedrotti M, Barbaro V, Di Iorio E, Ferrari S et al (2011) Long-term effectiveness of autologous cultured limbal stem cell grafts in patients with limbal stem cell deficiency due to chemical burns. Clin Experiment Ophthalmol 40:255–267
14. Basu S, Mohamed A, Chaurasia S, Sejpal K, Vemuganti GK, Sangwan VS (2011) Clinical outcomes of penetrating keratoplasty after autologous cultivated limbal epithelial transplantation for ocular surface burns. Am J Ophthalmol 152:917–924, e1
15. Nakamura T, Sotozono C, Bentley AJ, Mano S, Inatomi T, Koizumi N et al (2010) Long-term phenotypic study after allogeneic cultivated corneal limbal epithelial transplantation for severe ocular surface diseases. Ophthalmology 117:2247–2254, e1
16. Tseng SC, Meller D, Anderson DF, Touhami A, Pires RT, Gruterich M et al (2002) Ex vivo preservation and expansion of human limbal epithelial stem cells on amniotic membrane for treating corneal diseases with total limbal stem cell deficiency. Adv Exp Med Biol 506(Pt B):1323–1334
17. Gomes JA, Pazos HS, Silva AB, Cristovam PC, Belfort Junior R (2009) Transplante de celulas-tronco epiteliais limbicas alogenas expandidas ex vivo sobre membrana amniotica: relato de caso. Arq Bras Oftalmol 72:254–256
18. Harkin DG, Barnard Z, Gillies P, Ainscough SL, Apel AJ (2004) Analysis of p63 and cytokeratin expression in a cultivated limbal autograft used in the treatment of limbal stem cell deficiency. Br J Ophthalmol 88:1154–1158
19. Shigeyasu C, Shimazaki J (2011) Ocular surface reconstruction after exposure to high concentrations of antiseptic solutions. Cornea 31:59–65
20. Sangwan VS, Basu S, Vemuganti GK, Sejpal K, Subramaniam SV, Bandyopadhyay S et al (2011) Clinical outcomes of xeno-free autologous cultivated limbal epithelial transplantation: a 10-year study. Br J Ophthalmol 95:1525–1529
21. Sharma S, Tandon R, Mohanty S, Sharma NMV, Sen S, Kashyap S et al (2011) Culture of corneal limbal epithelial stem cells: experience from benchtop to bedside in a tertiary care hospital in India. Cornea 30:1223–1232
22. Thanos M, Pauklin M, Steuhl KP, Meller D (2010) Ocular surface reconstruction with cultivated limbal epithelium in a patient with unilateral stem cell deficiency caused by Epidermolysis Bullosa Dystrophica Hallopeau-Siemens. Cornea 29:462–464
23. Meller D, Pauklin M, Westekemper H, Steuhl KP (2010) Autologe Transplantation von kultiviertem Limbusepithel. Ophthalmologe 107:1133–1138
24. Desousa JL, Daya S, Malhotra R (2009) Adnexal surgery in patients undergoing ocular

surface stem cell transplantation. Ophthalmology 116:235–242

25. Meller D, Fuchsluger T, Pauklin M, Steuhl KP (2009) Ocular surface reconstruction in graft-versus-host disease with HLA-identical living-related allogeneic cultivated limbal epithelium after hematopoietic stem cell transplantation from the same donor. Cornea 28:233–236
26. Sangwan VS, Vemuganti GK, Singh S, Balasubramanian D (2003) Successful reconstruction of damaged ocular outer surface in humans using limbal and conjuctival stem cell culture methods. Biosci Rep 23:169–174
27. Ang LP, Sotozono C, Koizumi N, Suzuki T, Inatomi T, Kinoshita S (2007) A comparison between cultivated and conventional limbal stem cell transplantation for Stevens-Johnson syndrome. Am J Ophthalmol 143:178–180
28. Baradaran-Rafii A, Ebrahimi M, Kanavi MR, Taghi-Abadi E, Aghdami N, Eslani M et al (2010) Midterm outcomes of autologous cultivated limbal stem cell transplantation with or without penetrating keratoplasty. Cornea 29:502–509
29. Daya SM, Watson A, Sharpe JR, Giledi O, Rowe A, Martin R et al (2005) Outcomes and DNA analysis of ex vivo expanded stem cell allograft for ocular surface reconstruction. Ophthalmology 112:470–477
30. Di Girolamo N, Bosch M, Zamora K, Coroneo MT, Wakefield D, Watson SL (2009) A contact lens-based technique for expansion and transplantation of autologous epithelial progenitors for ocular surface reconstruction. Transplantation 87:1571–1578
31. Di Iorio E, Ferrari S, Fasolo A, Bohm E, Ponzin D, Barbaro V (2010) Techniques for culture and assessment of limbal stem cell grafts. Ocul Surf 8:146–153
32. Fatima A, Vemuganti GK, Iftekhar G, Rao GN, Sangwan VS (2007) In vivo survival and stratification of cultured limbal epithelium. Clin Experiment Ophthalmol 35:96–98
33. Grueterich M (2002) Phenotypic study of a case with successful transplantation of ex vivo expanded human limbal epithelium for unilateral total limbal stem cell deficiency. Ophthalmology 109:1547–1552
34. Kawashima M, Kawakita T, Satake Y, Higa K, Shimazaki J (2007) Phenotypic study after cultivated limbal epithelial transplantation for limbal stem cell deficiency. Arch Ophthalmol 125:1337–1344
35. Koizumi N, Inatomi T, Suzuki T, Sotozono C, Kinoshita S (2001) Cultivated corneal epithelial transplantation for ocular surface reconstruction in acute phase of Stevens-Johnson syndrome. Arch Ophthalmol 119:298–300
36. Nakamura T, Koizumi N, Tsuzuki M, Inoki K, Sano Y, Sotozono C et al (2003) Successful regrafting of cultivated corneal epithelium using amniotic membrane as a carrier in severe ocular surface disease. Cornea 22:70–71
37. Nakamura T, Inatomi T, Sotozono C, Koizumi N, Kinoshita S (2004) Successful primary culture and autologous transplantation of corneal limbal epithelial cells from minimal biopsy for unilateral severe ocular surface disease. Acta Ophthalmol Scand 82:468–471
38. Nakamura T, Inatomi T, Sotozono C, Ang LP, Koizumi N, Yokoi N et al (2006) Transplantation of autologous serum-derived cultivated corneal epithelial equivalents for the treatment of severe ocular surface disease. Ophthalmology 113:1765–1772
39. Pauklin M, Fuchsluger TA, Westekemper H, Steuhl KP, Meller D (2010) Midterm results of cultivated autologous and allogeneic limbal epithelial transplantation in limbal stem cell deficiency. Dev Ophthalmol 45:57–70
40. Rama P, Bonini S, Lambiase A, Golisano O, Paterna P, De Luca M et al (2001) Autologous fibrin-cultured limbal stem cells permanently restore the corneal surface of patients with total limbal stem cell deficiency. Transplantation 72:1478–1485
41. Sangwan VS, Vemuganti GK, Iftekhar G, Bansal AK, Rao GN (2003) Use of autologous cultured limbal and conjunctival epithelium in a patient with severe bilateral ocular surface disease induced by acid injury: a case report of unique application. Cornea 22:478–481
42. Sangwan VS, Matalia HP, Vemuganti GK, Fatima A, Iftekhar G, Singh S et al (2006) Clinical outcome of autologous cultivated limbal epithelium transplantation. Indian J Ophthalmol 54:29–34
43. Schwab IR, Reyes M, Isseroff RR (2000) Successful transplantation of bioengineered tissue replacements in patients with ocular surface disease. Cornea 19:421–426
44. Shimazaki J, Higa K, Morito F, Dogru M, Kawakita T, Satake Y et al (2007) Factors influencing outcomes in cultivated limbal epithelial transplantation for chronic cicatricial ocular surface disorders. Am J Ophthalmol 143:945–953
45. Tsai RJ, Li LM, Chen JK (2000) Reconstruction of damaged corneas by transplantation of autologous limbal epithelial cells. N Engl J Med 343:86–93
46. Colabelli Gisoldi RA, Pocobelli A, Villani CM, Amato D, Pellegrini G (2010) Evaluation

of molecular markers in corneal regeneration by means of autologous cultures of limbal cells and keratoplasty. Cornea 29:715–722

47. Kolli S, Ahmad S, Lako M, Figueiredo F (2010) Successful clinical implementation of corneal epithelial stem cell therapy for treatment of unilateral limbal stem cell deficiency. Stem Cells 28:597–610
48. Rama P, Matuska S, Paganoni G, Spinelli A, De Luca M, Pellegrini G (2010) Limbal stem-cell therapy and long-term corneal regeneration. N Engl J Med 363:147–155
49. Shortt AJ, Secker GA, Rajan MS, Meligonis G, Dart JK, Tuft SJ et al (2008) Ex vivo expansion and transplantation of limbal epithelial stem cells. Ophthalmology 115:1989–1997
50. Pan Z, Zhang W, Wu Y, Sun B (2002) Transplantation of corneal stem cells cultured on amniotic membrane for corneal burn: experimental and clinical study. Chin Med J (Engl) 115:767–769
51. Shimazaki J, Aiba M, Goto E, Kato N, Shimmura S, Tsubota K (2002) Transplantation of human limbal epithelium cultivated on amniotic membrane for the treatment of severe ocular surface disorders. Ophthalmology 109:1285–1290
52. Sangwan VS, Matalia HP, Vemuganti GK, Iftekhar G, Fatima A, Singh S et al (2005) Early results of penetrating keratoplasty after cultivated limbal epithelium transplantation. Arch Ophthalmol 123:334–340
53. Sangwan VS, Murphy SI, Vemuganti GK, Bansal AK, Gangopadhyay N, Rao GN (2005) Cultivated corneal epithelial transplantation for severe ocular surface disease in vernal keratoconjunctivitis. Cornea 24:426–430
54. Dobrowolski D, Wylegala E, Orzechowska-Wylegala B, Wowra B, Wroblewska-Czajka E (2011) Application of autologous cultivated corneal epithelium for corneal limbal stem cell insufficiency—short-term results. Klin Oczna 113:346–351
55. Fatima A, Matalia HP, Vemuganti GK, Honavar SG, Sangwan VS (2006) Pseudoepitheliomatous hyperplasia mimicking ocular surface squamous neoplasia following cultivated limbal epithelium transplantation. Clin Experiment Ophthalmol 34:889–891
56. Tsai RJ, Li L, Chen J (2000) Reconstruction of damaged corneas by transplantation of autologous limbal epithelial cells(1). Am J Ophthalmol 130:543
57. Shortt AJ, Secker GA, Notara MD, Limb GA, Khaw PT, Tuft SJ et al (2007) Transplantation of ex vivo cultured limbal epithelial stem cells: a review of techniques and clinical results. Surv Ophthalmol 52:483–502
58. Satake Y, Shimmura S, Shimazaki J (2009) Cultivated autologous limbal epithelial transplantation for symptomatic bullous keratopathy. BMJ Case Rep. pii: bcr.11.2008.1239
59. Mariappan I, Maddileti S, Savy S, Tiwari S, Gaddipati S, Fatima A et al (2010) In vitro culture and expansion of human limbal epithelial cells. Nat Protoc 5:1470–1479
60. Madhavan HN, Priya K, Malathi J, Joseph PR (2002) Preparation of amniotic membrane for ocular surface reconstruction. Indian J Ophthalmol 50:227–231
61. Dua HS, Maharajan VS, Hopkinson A (2006) Controversies and limitations of amniotic membrane in ophthalmic surgery. In: Reinhard T, Larkin F (eds) Cornea and external eye disease (Essentials in ophthalmology). Springer, Berlin, Heidelberg, pp 21–33
62. Adds PJ, Hunt CJ, Dart JK (2001) Amniotic membrane grafts, "fresh" or frozen? A clinical and in vitro comparison. Br J Ophthalmol 85:905–907
63. Lee SH, Tseng SC (1997) Amniotic membrane transplantation for persistent epithelial defects with ulceration. Am J Ophthalmol 123:303–312
64. van Baare J, Ligtvoet EE, Middelkoop E (1998) Microbiological evaluation of glycerolized cadaveric donor skin. Transplantation 65:966–970
65. Shortt AJ, Secker GA, Lomas RJ, Wilshaw SP, Kearney JN, Tuft SJ et al (2009) The effect of amniotic membrane preparation method on its ability to serve as a substrate for the ex-vivo expansion of limbal epithelial cells. Biomaterials 30:1056–1065
66. Grueterich M, Espana EM, Tseng SC (2003) Modulation of keratin and connexin expression in limbal epithelium expanded on denuded amniotic membrane with and without a 3T3 fibroblast feeder layer. Invest Ophthalmol Vis Sci 44:4230–4236
67. Grueterich M, Espana E, Tseng SC (2002) Connexin 43 expression and proliferation of human limbal epithelium on intact and denuded amniotic membrane. Invest Ophthalmol Vis Sci 43:63–71
68. Sudha B, Sitalakshmi G, Iyer GK, Krishnakumar S (2008) Putative stem cell markers in limbal epithelial cells cultured on intact & denuded human amniotic membrane. Indian J Med Res 128:149–156
69. Koizumi N, Rigby H, Fullwood NJ, Kawasaki S, Tanioka H, Koizumi K et al (2007) Comparison of intact and denuded amniotic membrane as a substrate for cell-suspension

culture of human limbal epithelial cells. Graefes Arch Clin Exp Ophthalmol 245: 123–134

70. Kubo M, Sonoda Y, Muramatsu R, Usui M (2001) Immunogenicity of human amniotic membrane in experimental xenotransplantation. Invest Ophthalmol Vis Sci 42: 1539–1546
71. Dekaris I, Gabric N (2009) Preparation and preservation of amniotic membrane. Dev Ophthalmol 43:97–104
72. Thomasen H, Pauklin M, Noelle B, Geerling G, Vetter J, Steven P et al (2011) The effect of long-term storage on the biological and histological properties of cryopreserved amniotic membrane. Curr Eye Res 36:247–255
73. Meller D, Pires RT, Tseng SC (2002) Ex vivo preservation and expansion of human limbal epithelial stem cells on amniotic membrane cultures. Br J Ophthalmol 86:463–471
74. Tseng SC, Li DQ, Ma X (1999) Suppression of transforming growth factor-beta isoforms, TGF-beta receptor type II, and myofibroblast differentiation in cultured human corneal and limbal fibroblasts by amniotic membrane matrix. J Cell Physiol 179:325–335
75. Grueterich M, Tseng SC (2002) Human limbal progenitor cells expanded on intact amniotic membrane ex vivo. Arch Ophthalmol 120:783–790
76. Hernandez Galindo EE, Galindo EE, Theiss C, Steuhl KP, Meller D (2003) Gap junctional communication in microinjected human limbal and peripheral corneal epithelial cells cultured on intact amniotic membrane. Exp Eye Res 76:303–314
77. Hernandez Galindo EE, Theiss C, Steuhl KP, Meller D (2003) Expression of Delta Np63 in response to phorbol ester in human limbal epithelial cells expanded on intact human amniotic membrane. Invest Ophthalmol Vis Sci 44:2959–2965
78. Du Y, Chen J, Funderburgh JL, Zhu X, Li L (2003) Functional reconstruction of rabbit corneal epithelium by human limbal cells cultured on amniotic membrane. Mol Vis 9: 635–643
79. Ma DH, Lai JY, Cheng HY, Tsai CC, Yeh LK (2010) Carbodiimide cross-linked amniotic membranes for cultivation of limbal epithelial cells. Biomaterials 31:6647–6658
80. Lim LS, Poh RW, Riau AK, Beuerman RW, Tan D, Mehta JS (2010) Biological and ultrastructural properties of acelagraft, a freeze-dried gamma-irradiated human amniotic membrane. Arch Ophthalmol 128:1303–1310
81. Singh R, Gupta P, Kumar P, Kumar A, Chacharkar MP (2003) Properties of air dried radiation processed amniotic membranes under different storage conditions. Cell Tissue Bank 4:95–100
82. Thomasen H, Pauklin M, Steuhl KP, Meller D (2009) Comparison of cryopreserved and air-dried human amniotic membrane for ophthalmologic applications. Graefes Arch Clin Exp Ophthalmol 247:1691–1700
83. Connon CJ, Doutch J, Chen B, Hopkinson A, Mehta JS, Nakamura T et al (2010) The variation in transparency of amniotic membrane used in ocular surface regeneration. Br J Ophthalmol 94:1057–1061
84. Rahman I, Said DG, Maharajan VS, Dua HS (2009) Amniotic membrane in ophthalmology: indications and limitations. Eye 23:1954–1961
85. Hopkinson A, McIntosh RS, Tighe PJ, James DK, Dua HS (2006) Amniotic membrane for ocular surface reconstruction: donor variations and the effect of handling on TGF-beta content. Invest Ophthalmol Vis Sci 47:4316–4322
86. Talbot M, Carrier P, Giasson CJ, Deschambeault A, Guerin SL, Auger FA et al (2006) Autologous transplantation of rabbit limbal epithelia cultured on fibrin gels for ocular surface reconstruction. Mol Vis 12:65–75
87. Higa K, Takeshima N, Moro F, Kawakita T, Kawashima M, Demura M et al (2010) Porous silk fibroin film as a transparent carrier for cultivated corneal epithelial sheets. J Biomater Sci Polym Ed. doi:10.1163/092050610X538218
88. Bray LJ, George KA, Ainscough SL, Hutmacher DW, Chirila TV, Harkin DG (2011) Human corneal epithelial equivalents constructed on Bombyx mori silk fibroin membranes. Biomaterials 32:5086–5091
89. Mi S, Chen B, Wright B, Connon CJ (2010) Ex vivo construction of an artificial ocular surface by combination of corneal limbal epithelial cells and a compressed collagen scaffold containing keratocytes. Tissue Eng Part A 16:2091–2100
90. Levis HJ, Brown RA, Daniels JT (2010) Plastic compressed collagen as a biomimetic substrate for human limbal epithelial cell culture. Biomaterials 31:7726–7737
91. Mi S, Chen B, Wright B, Connon CJ (2010) Plastic compression of a collagen gel forms a much improved scaffold for ocular surface tissue engineering over conventional collagen gels. J Biomed Mater Res A 95: 447–453

92. Barbaro V, Ferrari S, Fasolo A, Ponzin D, Di Iorio E (2009) Reconstruction of a human hemicornea through natural scaffolds compatible with the growth of corneal epithelial stem cells and stromal keratocytes. Mol Vis 15:2084–2093
93. Zajicova A, Pokorna K, Lencova A, Krulova M, Svobodova E, Kubinova S et al (2010) Treatment of ocular surface injuries by limbal and mesenchymal stem cells growing on nanofiber scaffolds. Cell Transplant 19:1281–1290
94. Deshpande P, McKean R, Blackwood KA, Senior RA, Ogunbanjo A, Ryan AJ et al (2010) Using poly(lactide-co-glycolide) electrospun scaffolds to deliver cultured epithelial cells to the cornea. Regen Med 5:395–401
95. Sharma S, Mohanty S, Gupta D, Jassal M, Agrawal AK, Tandon R (2011) Cellular response of limbal epithelial cells on electrospun poly-epsilon-caprolactone nanofibrous scaffolds for ocular surface bioengineering: a preliminary in vitro study. Mol Vis 17:2898–2910
96. Fiorica C, Senior RA, Pitarresi G, Palumbo FS, Giammona G, Deshpande P et al (2011) Biocompatible hydrogels based on hyaluronic acid cross-linked with a polyaspartamide derivative as delivery systems for epithelial limbal cells. Int J Pharm 414:104–111
97. Di Girolamo N, Chui J, Wakefield D, Coroneo MT (2007) Cultured human ocular surface epithelium on therapeutic contact lenses. Br J Ophthalmol 91:459–464
98. Li DQ, Chen Z, Song XJ, de Paiva CS, Kim HS, Pflugfelder SC (2005) Partial enrichment of a population of human limbal epithelial cells with putative stem cell properties based on collagen type IV adhesiveness. Exp Eye Res 80:581–590
99. Kito K, Kagami H, Kobayashi C, Ueda M, Terasaki H (2005) Effects of cryopreservation on histology and viability of cultured corneal epithelial cell sheets in rabbit. Cornea 24:735–741
100. Ahmadiankia N, Ebrahimi M, Hosseini A, Baharvand H (2009) Effects of different extracellular matrices and co-cultures on human limbal stem cell expansion in vitro. Cell Biol Int 33:978–987
101. Sudha B, Jasty S, Krishnan S, Krishnakumar S (2009) Signal transduction pathway involved in the ex vivo expansion of limbal epithelial cells cultured on various substrates. Indian J Med Res 129:382–389
102. Nishida K, Yamato M, Hayashida Y, Watanabe K, Yamamoto K, Adachi E et al (2004) Corneal reconstruction with tissue-engineered cell sheets composed of autologous oral mucosal epithelium. N Engl J Med 351: 1187–1196
103. Nishida K, Yamato M, Hayashida Y, Watanabe K, Maeda N, Watanabe H et al (2004) Functional bioengineered corneal epithelial sheet grafts from corneal stem cells expanded ex vivo on a temperature-responsive cell culture surface. Transplantation 77:379–385
104. Hayashi R, Yamato M, Takayanagi H, Oie Y, Kubota A, Hori Y et al (2010) Validation system of tissue engineered epithelial cell sheets for corneal regenerative medicine. Tissue Eng Part C Methods 16:553–560
105. Sudha B, Madhavan HN, Sitalakshmi G, Malathi J, Krishnakumar S, Mori Y et al (2006) Cultivation of human corneal limbal stem cells in Mebiol gel—A thermo-reversible gelation polymer. Indian J Med Res 124:655–664
106. Sitalakshmi G, Sudha B, Madhavan HN, Vinay S, Krishnakumar S, Mori Y et al (2008) Ex vivo cultivation of corneal limbal epithelial cells in a thermoreversible polymer (Mebiol gel) and their transplantation in rabbits: an animal model. Tissue Eng Part A 75:402–415
107. Francis D, Abberton K, Thompson E, Daniell M (2009) Myogel supports the ex-vivo amplification of corneal epithelial cells. Exp Eye Res 88:339–346
108. Xie HT, Chen SY, Li GG, Tseng SC (2011) Limbal epithelial stem/progenitor cells attract stromal niche cells by SDF-1/CXCR4 signaling to prevent differentiation. Stem Cells 29:1874–1885
109. Dravida S, Gaddipati S, Griffith M, Merrett K, Lakshmi Madhira S, Sangwan VS et al (2008) A biomimetic scaffold for culturing limbal stem cells: a promising alternative for clinical transplantation. J Tissue Eng Regen Med 2:263–271
110. Utheim TP, Raeder S, Utheim OA, Cai Y, Roald B, Drolsum L et al (2007) A novel method for preserving cultured limbal epithelial cells. Br J Ophthalmol 91:797–800
111. Raeder S, Utheim TP, Utheim OA, Cai Y, Roald B, Lyberg T et al (2007) Effect of limbal explant orientation on the histology, phenotype, ultrastructure and barrier function of cultured limbal epithelial cells. Acta Ophthalmol Scand 85:377–386
112. Meller D, Tseng SC (1999) Conjunctival epithelial cell differentiation on amniotic membrane. Invest Ophthalmol Vis Sci 40: 878–886
113. James SE, Rowe A, Ilari L, Daya S, Martin R (2001) The potential for eye bank limbal rings to generate cultured corneal epithelial allografts. Cornea 20:488–494

114. Meyer-Blazejewska EA, Kruse FE, Bitterer K, Meyer C, Hofmann-Rummelt C, Wunsch PH (2010) Preservation of the limbal stem cell phenotype by appropriate culture techniques. Invest Ophthalmol Vis Sci 51:765–774

115. Notara M, Shortt AJ, O'Callaghan AR, Daniels JT (2012) The impact of age on the physical and cellular properties of the human limbal stem cell niche. Age (Dordr). doi:10.1007/s11357-011-9359-5

116. Chang CY, McGhee JJ, Green CR, Sherwin T (2011) Comparison of stem cell properties in cell populations isolated from human central and limbal corneal epithelium. Cornea 30:1155–1162

117. Zito-Abbad E, Borderie VM, Baudrimont M, Bourcier T, Laroche L, Chapel C et al (2006) Corneal epithelial cultures generated from organ-cultured limbal tissue: factors influencing epithelial cell growth. Curr Eye Res 31:391–399

118. Shanmuganathan VA, Rotchford AP, Tullo AB, Joseph A, Zambrano I, Dua HS (2006) Epithelial proliferative potential of organ cultured corneoscleral rims; implications for allolimbal transplantation and eye banking. Br J Ophthalmol 90:55–58

119. Kim HS, Jun Song X, de Paiva CS, Chen Z, Pflugfelder SC, Li DQ (2004) Phenotypic characterization of human corneal epithelial cells expanded ex vivo from limbal explant and single cell cultures. Exp Eye Res 79:41–49

120. Utheim TP, Raeder S, Olstad OK, Utheim OA, de LaPaz M, Cheng R et al (2009) Comparison of the histology, gene expression profile, and phenotype of cultured human limbal epithelial cells from different limbal regions. Invest Ophthalmol Vis Sci 50:5165–5172

121. Shortt AJ, Secker GA, Munro PM, Khaw PT, Tuft SJ, Daniels JT (2007) Characterization of the limbal epithelial stem cell niche: novel imaging techniques permit in vivo observation and targeted biopsy of limbal epithelial stem cells. Stem Cells 25:1402–1409

122. Pellegrini G, Golisano O, Paterna P, Lambiase A, Bonini S, Rama P et al (1999) Location and clonal analysis of stem cells and their differentiated progeny in the human ocular surface. J Cell Biol 145:769–782

123. Espana EM, Romano AC, Kawakita T, Di Pascuale M, Smiddy R, Tseng SC (2003) Novel enzymatic isolation of an entire viable human limbal epithelial sheet. Invest Ophthalmol Vis Sci 44:4275–4281

124. Chen SY, Hayashida Y, Chen MY, Xie HT, Tseng SC (2010) A new isolation method of human limbal progenitor cells by maintaining close association with their niche cells. Tissue Eng Part C Methods 17:537–548

125. Kawakita T, Shimmura S, Higa K, Espana EM, He H, Shimazaki J et al (2009) Greater growth potential of p63-positive epithelial cell clusters maintained in human limbal epithelial sheets. Invest Ophthalmol Vis Sci 50:4611–4617

126. Cristovam PC, Gloria MA, Melo GB, Gomes JA (2008) Importancia do co-cultivo com fibroblastos de camundongo 3T3 para estabelecer cultura de suspensao de celulas epiteliais do limbo humano. Arq Bras Oftalmol 71:689–694

127. Schwab IR, Johnson NT, Harkin DG (2006) Inherent risks associated with manufacture of bioengineered ocular surface tissue. Arch Ophthalmol 124:1734–1740

128. Selver OB, Barash A, Ahmed M, Wolosin JM (2011) ABCG2-dependent dye exclusion activity and clonal potential in epithelial cells continuously growing for one month from limbal explants. Invest Ophthalmol Vis Sci 52:4330–4337

129. Schofield R (1983) The stem cell system. Biomed Pharmacother 37:375–380

130. Watt FM (2000) Out of Eden: stem cells and their niches. Science 287:1427–1430

131. Joseph A, Powell-Richards AO, Shanmuganathan VA, Dua HS (2004) Epithelial cell characteristics of cultured human limbal explants. Br J Ophthalmol 88:393–398

132. Kolli S, Lako M, Figueiredo F, Mudhar H, Ahmad S (2008) Loss of corneal epithelial stem cell properties in outgrowths from human limbal explants cultured on intact amniotic membrane. Regen Med 3:329–342

133. Zhang X, Sun H, Tang X, Ji J, Li X, Sun J et al (2005) Comparison of cell-suspension and explant culture of rabbit limbal epithelial cells. Exp Eye Res 80:227–233

134. Tan EK, He H, Tseng SC (2011) Epidermal differentiation and loss of clonal growth potential of human limbal basal epithelial progenitor cells during intrastromal invasion. Invest Ophthalmol Vis Sci 52:4534–4545

135. Li W, Hayashida Y, He H, Kuo CL, Tseng SC (2007) The fate of limbal epithelial progenitor cells during explant culture on intact amniotic membrane. Invest Ophthalmol Vis Sci 48:605–613

136. Koizumi N, Cooper LJ, Fullwood NJ, Nakamura T, Inoki K, Tsuzuki M et al (2002) An evaluation of cultivated corneal limbal epithelial cells, using cell-suspension culture. Invest Ophthalmol Vis Sci 43:2114–2121

137. Deshpande P, Notara M, Bullett N, Daniels JT, Haddow DB, MacNeil S (2009) Development of a surface modified contact lens for transfer of cultured limbal epithelial cells for ocular surface diseases. Tissue Eng Part A 15:2889–2902

138. Raeder S et al (2008) Effects of the size of explants on adhesion and outgrowth of ex vivo expanded limbal epithelial cells. Invest Ophthalmol Vis Sci 47, E-Abstract 5717

139. Steele JG, Johnson G, Griesser HJ, Underwood PA (1997) Mechanism of initial attachment of corneal epithelial cells to polymeric surfaces. Biomaterials 18:1541–1551

140. Liu L, Hartwig D, Harloff S, Herminghaus P, Wedel T, Geerling G (2005) An optimised protocol for the production of autologous serum eyedrops. Graefes Arch Clin Exp Ophthalmol 243:706–714

141. Thomas L (1998) Labor und diagnose. indikation und Bewertung von laborbefunden für die medizinische diagnostik, 5th edn. TH-Books, Frankfurt/Main, p 1494

142. Dugrillon A et al (2002) Platelets applied to wounds and in tissue regeneration: induction of proliferation apoptosis by platelet membranes. In: Mempel W, Schramm W (eds) Infusion therapy and transfusion medicine. Kluwer Academic, Amsterdam, pp 70–71

143. Geerling G, Hartwig D (2005) Autologous serum eye drops for ocular surface disorders. In: Reinhard T, Larkin DFP (eds) Cornea and external eye disease. Springer, Berlin, pp 2–18

144. Geerling G, Maclennan S, Hartwig D (2004) Autologous serum eye drops for ocular surface disorders. Br J Ophthalmol 88:1467–1474

145. Green H, Kehinde O, Thomas J (1979) Growth of cultured human epidermal cells into multiple epithelia suitable for grafting. Proc Natl Acad Sci U S A 76:5665–5668

146. Nakamura T, Ang LP, Rigby H, Sekiyama E, Inatomi T, Sotozono C et al (2006) The use of autologous serum in the development of corneal and oral epithelial equivalents in patients with Stevens-Johnson syndrome. Invest Ophthalmol Vis Sci 47:909–916

147. Shahdadfar A, Haug K, Pathak M, Drolsum L, Olstad OK, Johnsen EO et al (2012) Ex Vivo expanded autologous limbal epithelial cells on amniotic membrane using a culture medium with human serum as single supplement. Exp Eye Res 97:1–9

148. Ban Y, Cooper LJ, Fullwood NJ, Tsuzuki M, Koizumi N, Dota A et al (2003) Comparison of ultrastructure, tight junction-related protein expression and barrier function of human corneal epithelial cells cultivated on amniotic membrane with and without air-lifting. Exp Eye Res 76:735–743

149. Ghoubay-Benallaoua D, Basli E, Goldschmidt P, Pecha F, Chaumail C, Laroche L et al (2011) Human epithelial cell cultures from superficial limbal explants. Mol Vis 17:341–354

150. D'Ippolito G, Diabira S, Howard GA, Roos BA, Schiller PC (2006) Low oxygen tension inhibits osteogenic differentiation and enhances stemness of human MIAMI cells. Bone 39:513–522

151. Cipolleschi MG, Dello SP, Olivotto M (1993) The role of hypoxia in the maintenance of hematopoietic stem cells. Blood 82: 2031–2037

152. Li C, Yin T, Dong N, Dong F, Fang X, Qu YL et al (2011) Oxygen tension affects terminal differentiation of corneal limbal epithelial cells. J Cell Physiol 226:2429–2437

153. Miyashita H, Higa K, Kato N, Kawakita T, Yoshida S, Tsubota K et al (2007) Hypoxia enhances the expansion of human limbal epithelial progenitor cells in vitro. Invest Ophthalmol Vis Sci 48:3586–3593

154. Higa K, Shimazaki J (2008) Recent advances in cultivated epithelial transplantation. Cornea 27:S41–S47

155. O'Callaghan AR, Daniels JT, Mason C (2011) Effect of sub-atmospheric oxygen on the culture of rabbit limbal epithelial cells. Curr Eye Res 36:691–698

156. Zakaria N, Koppen C, Van Tendeloo V, Berneman Z, Hopkinson A, Tassignon MJ (2010) Standardized limbal epithelial stem cell graft generation and transplantation. Tissue Eng Part C Methods 16:921–927

157. Kawakita T, Espana EM, He H, Li W, Liu CY, Tseng SC (2005) Intrastromal invasion by limbal epithelial cells is mediated by epithelial-mesenchymal transition activated by air exposure. Am J Pathol 167:381–393

158. Chen B, Mi S, Wright B, Connon CJ (2010) Differentiation status of limbal epithelial cells cultured on intact and denuded amniotic membrane before and after air-lifting. Tissue Eng Part A 16:2721–2729

159. Li W, Hayashida Y, Chen YT, He H, Tseng DY, Alonso M et al (2008) Air exposure induced squamous metaplasia of human limbal epithelium. Invest Ophthalmol Vis Sci 49:154–162

160. Kishida H, Kuroiwa Y (2007) Prevention of prion disease transmission. Nippon Rinsho 65:1454–1459

161. Raeder S, Utheim TP, Messelt E, Lyberg T (2010) The impact of de-epithelialization of the amniotic membrane matrix on morphology

of cultured human limbal epithelial cells subject to eye bank storage. Cornea 29:439–445

162. Utheim TP, Raeder S, Utheim OA, de la Paz M, Raold B, Lyberg T (2009) Sterility control and long-term eye-bank storage of cultured human limbal epithelial cells for transplantation. Br J Ophthalmol 93: 980–983
163. Raeder S, Utheim TP, Utheim OA, Nicolaissen B, Raold B, Cai Y et al (2007) Effects of organ culture and optisol-GS storage on structural integrity, phenotypes, and apoptosis in cultured corneal epithelium. Invest Ophthalmol Vis Sci 48:5484–5493
164. Utheim TP et al (2009) Storage of cultured human limbal epithelial cells in Optisol-GS at 23 °C versus 5 °C. Invest Ophthalmol Vis Sci 50, E-Abstract 1778
165. Raeder S et al (2009) Genome-wide transcriptional analysis of cultured human limbal epithelial cells following hypothermic storage Optisol-GS. Invest Ophthalmol Vis Sci 50, E-Abstract 1773/A441
166. Utheim TP et al (2011) Comparison of storage of cultured human limbal epithelial cells in HEPES-MEM, Optisol-GS and PAA Quantum 286 for 4 days at 23 °C. Invest Ophthalmol Vis Sci 52, E-Abstract 5124
167. Utheim TP et al (2012) Transportation simulations of cultured human limbal epithelial cells subjected to eye-bank storage. Invest Ophthalmol Vis Sci 53, E-Abstract 183

Chapter 8

The Culture of Limbal Stromal Cells and Corneal Endothelial Cells

Naresh Polisetti and Nancy C. Joyce

Abstract

The cornea is the transparent front part of the eye and comprises three distinct cell layers. One of these cell layers is a self-renewing epithelium long believed to harbor a resident stem cell population. The location and characteristics of corneal epithelial stem cells have now been confirmed by several research groups, and these cells are currently applied therapeutically. The corneal stroma and endothelium are largely quiescent after infancy, and until recently they were not considered to undergo self-renewal or to maintain stem cells. This view was overturned during the last two decades. At present, cell populations with characteristics of adult stem cells are routinely isolated and characterized from the limbal stroma and the corneal endothelium. This chapter describes methods for isolation and culture of limbal stromal cells and corneal endothelial cells.

Key words Limbal stromal stem cells, Corneal endothelial cells, Corneal stroma, Corneal endothelium, Limbal stromal cell culture, Corneal endothelial cell culture

1 Introduction

The cornea is the main structure that controls and focuses light entering the eye to the retina, to enable vision. Hence, its transparency is critical for sight. The cornea is composed of an upper epithelial layer supported by a stromal layer, an endothelial layer, and two acellular layers (Bowman's membrane and Descemet's membrane) (*see* Chapter 1) (Fig. 1a) [1].

The corneal epithelium consists of 5–6 layers of stratified, nonkeratinized epithelial cells, responsible for protecting the eye from foreign material, including pathogens, as well as absorbing oxygen and nutrients from the tear film. Corneal integrity and function are dependent upon the self-renewing properties of the corneal epithelium, which is maintained by the presence of stem cells located in the limbal region, at the cornea–conjunctiva interface. Limbal epithelial stem cells (LESC) (Fig. 1b) can divide symmetrically to self-renew, and asymmetrically to produce daughter transient

Bernice Wright and Che J. Connon (eds.), *Corneal Regenerative Medicine: Methods and Protocols*, Methods in Molecular Biology, vol. 1014, DOI 10.1007/978-1-62703-432-6_8, © Springer Science+Business Media New York 2013

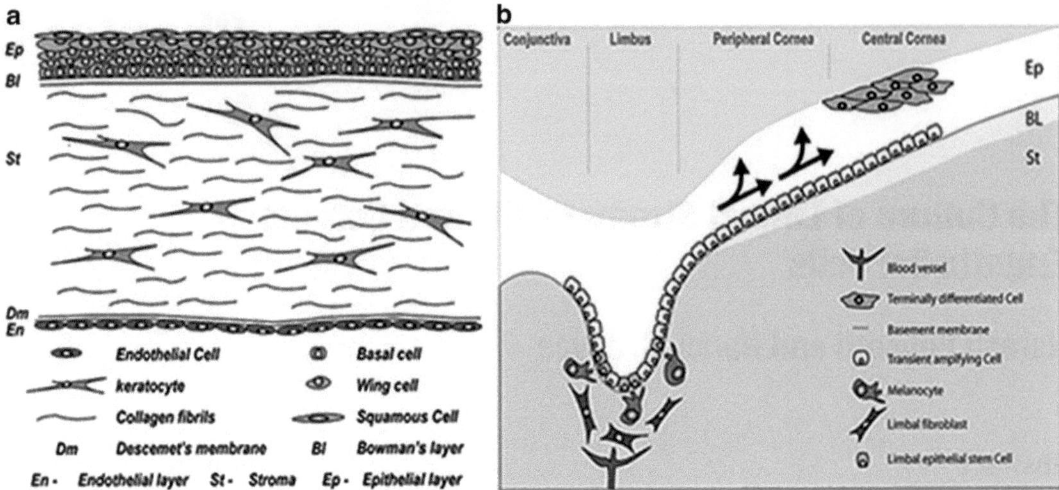

Fig. 1 (**a**) The human cornea in cross section. The outer surface comprises an epithelial layer that rests on a basement membrane, which in turn overlays a cell-free layer: Bowman's membrane. The middle cellular stromal layer contains a mainly collagenous extracellular matrix sparsely populated with keratocytes. The final layer consists of a single sheet of endothelial cells, which is physically separated from the stroma by the acellular Descemet's membrane. (**b**) Diagram of the human limbus. Limbal epithelial stem cells reside in the basal layer of the epithelium, at the limbus where the peripheral cornea meets the conjunctiva. Daughter transient amplifying cells divide and migrate towards the central cornea (*arrows*) to replenish the epithelium, which rests on Bowman's layer. The stroma, of the limbal epithelial stem cell niche, is populated with fibroblasts and melanocytes and also has a blood supply (From: Secker and Daniels (2009) [1])

amplifying cells (TAC) that migrate centripetally to populate the basal layer of the corneal epithelium [2, 3]. TAC divide and migrate superficially, progressively becoming more differentiated, and eventually becoming post-mitotic, terminally differentiated cells.

In the normal, uninjured cornea, LESC are mitotically quiescent and maintained in a specialized limbal stromal microenvironment or "niche." Upon corneal epithelial wounding, however, stem cells located in the limbus undergo self-renewal and proliferate to generate TAC which replace damaged epithelium. It is generally agreed that LESC are characterized by their location in the limbus, clonality (*see* Chapter 3), cytokeratin profile, and expression of transformation-related protein 63 (p63)Δ and ATP-binding cassette sub-family G member 2 (ABCG2) [4–6]. It is well established that the niche plays an important role in the maintenance of stem cell properties in several tissues, and this is expected to be true in the case of the LESC niche [7–10]. Some of the factors assumed necessary for niche regulation, include proximity to vasculature [11], the basement membrane composition with respect to specific isoforms of collagen IV, laminin, and fibronectin [12], and the presence of limbal keratocytes in the underlying stroma, which produce various cytokines [13]. There is evidence that the limbal niche also hosts stromal stem cells, with apparent multi-lineage differentiation potential [14, 15]. The phenotype of limbal stromal cells has been previously characterized by variable expression of several putative stem cell markers. The lack of agreement, however, on specific molecular markers for the identification of a pluripotent subpopulation among limbal stromal cells has to date limited the investigation of their differentiation potential to a small number of studies [16–18].

The single cell layer posterior endothelium is essential for the maintenance of appropriate stromal hydration to support corneal transparency. Corneal endothelial cells contain Na^+/K^+ ATPase pumps that help maintain stromal thickness [19]. The human corneal endothelium does not undergo functional regeneration in vivo [20–22]. Various laboratories [23–32] have demonstrated the capacity of human corneal endothelial cells (HCEC) to proliferate in vitro. The procedures involved in the isolation and subsequent cultivation of these primary HCEC vary greatly, however, amongst laboratories in their isolation protocols, extracellular matrices used to aid cell attachment, and most importantly, the media used in HCEC culture and propagation.

In this chapter, we describe protocols for in vitro expansion of limbal stromal cells and corneal endothelial cells using a small corneal tissue biopsy sample. The biopsy sample is either freshly harvested autologously from the healthy eye of an individual or allogenically from a live donor. Alternatively, tissue is obtained from a registered eye bank from the limbal rings of cadaveric donor corneas.

2 Materials

2.1 Isolation and Culture of Limbal Stromal Cells

1. Corneoscleral rims (*see* **Note 1**).
2. Tissue dissection kit: Sterile sharp-tipped, Beaver sclerotome, curve-tipped and blunt-ended forceps, and corneal scissors (11 cm (4-3/8 in.)).
3. Culture Medium: The culture medium should consist of Dulbecco's minimal essential medium (DMEM)/Ham's F12 solution containing 10 % (v/v) fetal calf serum (FCS), and 1× penicillin–streptomycin solution. This solution can be stored at 4 °C for up to 1 month.
4. Humidified, CO_2 incubator.
5. Phosphate buffered saline (PBS: 1× Solution): To prepare 1 L, dissolve PBS in 900 mL Milli-Q water according to the manufacturer's instructions and, adjust the pH to 7.2 with 1 N HCl or 1 N NaOH while stirring. Adjust the final volume to 1 L with Milli-Q water and sterilize by autoclaving. Store at 4 °C for up to 1 week.

6. Dispase II (50 mg/mL): Dissolve 500 mg dispase into 10 mL DMEM containing 10 % (v/v) FCS and filter-sterilize using 0.22 µm syringe filters. Prepare this solution immediately before use.
7. Collagenase IV (1 mg/mL): Dissolve 10 mg collagenase IV into 10 mL DMEM containing 10 % (w/v) FCS and filter-sterilize using a 0.22 µm syringe filter. Fresh solution to be prepared before use.
8. Trypsin–EDTA 0.05 % (Gibco).
9. Refrigerated centrifuge.
10. Inverted phase-contrast microscope.
11. Class II biosafety cabinet.
12. Water purification system and filtering units.
13. Sterile syringe filters and syringes.
14. Plastic dishes (tissue culture grade).
15. Sterile conical centrifuge tubes (50 or 15 mL).
16. Disposable plastic pipettes, pipette tips, and pipette aids.
17. Plastic spatula and cell lifter or cell scraper.
18. Cell strainer (70 µm nylon mesh).
19. Freezing solution: To prepare 10 mL of freezing solution, combine 5 mL DMEM solution, 4 mL FCS, and 1 mL dimethyl sulfoxide (DMSO). Mix thoroughly and filter-sterilize using a 0.22 µm syringe filter. Prepare solution on ice.

2.2 Isolation and Culture of Corneal Endothelial Cells

1. Corneas or corneoscleral rims.
2. Tissue dissection kit: Sterile sharp-tipped, Beaver sclerotome, curve-tipped and blunt-ended forceps, and corneal scissors (11 cm (4-3/8 in.)).
3. Dulbecco's PBS (Gibco).
4. Trypsin–EDTA 0.02–0.05 % (w/v) (Gibco).
5. Fetal bovine serum (FBS).
6. 0.02 % (w/v) Ethylenediaminetetraacetic acid (EDTA) solution.
7. FNC Coating mix—Coat plates following the manufacturer's instructions (Athena Enzyme Systems).
8. Culture media—OptiMEM-1 Minimal Essential Medium (MEM) containing 8 % (v/v) FBS, 5 ng/mL epidermal growth factor (EGF), 20 ng/mL nerve growth factor (NGF), 100 µg/mL pituitary extract, 20 µg/mL ascorbic acid, 200 µg/mL calcium chloride, 0.08 % (w/v) chondroitin sulfate, 50 µg/mL gentamycin, and 1× penicillin–streptomycin solution.
9. Humidified, CO_2 incubator.

10. Inverted phase-contrast microscope.
11. Class II biosafety cabinet.
12. Plastic dishes (tissue culture grade).
13. Flame polished glass pipette.

3 Methods

3.1 Isolation and Culture of Limbal Stromal Cells

For all procedures, approval from appropriate regulatory authorities and ethical committees is mandatory.

1. Obtain corneoscleral rims from cadaveric eyes or limbal biopsies from healthy donors with informed consent (*see* **Note 1**).
2. Rinse corneoscleral rims three times with DMEM containing 50 μg/mL gentamycin and 1.25 μg/mL amphotericin B.
3. Carefully remove excess sclera (leaving a scleral rim approximately 1 mm depth), conjunctiva, iris, and transparent cornea using a straight steel blade and a Beaver sclerotome. Ensure that sclera, transparent cornea and conjunctiva are completely removed (*see* **Note 2**).
4. Wash the isolated limbal rim with PBS in a culture dish.
5. Transfer the limbal rim into a well of a 6-well plate. Incubate the limbal rim in 3 mL dispase II in magnesium and calcium-free Hank's balanced salt solution for 2 h in a humidified incubator maintained at 37 °C with a 5 % CO_2 supply.
6. After the incubation period, remove the dispase solution and wash the limbal rim with 2 mL PBS.
7. Use a plastic spatula to scrape the epithelium side of the stroma. Observe through the microscope to ensure that epithelial layers are removed.
8. Wash the limbal stroma twice with 2 mL PBS.
9. Cut limbal segments into small pieces (explants: approximately 1 mm^2) and place explants into 3 mL collagenase solution (1 mg/mL) in a well of a 6-well plate, for 18 h in a humidified incubator maintained at 37 °C with a 5 % CO_2 supply (*see* **Note 3**).
10. After 8 h, agitate the sample with a 1 mL pipette tip several times, to release cells from the stroma (*see* **Note 4**).
11. Transfer the collagenase-digested sample to a 15 mL conical tube, wash the culture dish with 2 mL culture medium and add the wash solution to the collagenase solution.
12. Centrifuge the sample at 400 × g for 5 min and discard the supernatant.
13. Resuspend cells in 5 mL DMEM/F-12.

14. Pass the cell suspension through a 60 μm nylon filter to remove tissue remnants.
15. Centrifuge the filtered cell suspension at 300 × *g* for 5 min.
16. Remove supernatant from the cell pellet and resuspend cells with 1 mL culture medium.
17. Mix one part cell suspension (10 μL) with one part (10 μL) 0.4 % (w/v) Trypan blue dye.
18. Apply 10 μL Trypan blue/cell mixture to a hemocytometer. Place the hemocytometer on the stage of a phase-contrast microscope and focus on the cells.
19. Count unstained (viable) and stained (blue: nonviable) cells. Calculate the cell concentration per mL: average number of cells in one large square × dilution factor × 10^4. Calculate the viability of cells: percentage viable cells = total number of viable cells per mL of aliquot/total number of cells per mL of aliquot × 100.
20. Transfer 5×10^4 cells to a T25 culture flask and flood the flask with culture medium (approximately 4 mL).
21. Place the T25 flask in a humidified incubator maintained at 37 °C with a 5 % CO_2 supply.
22. After 24 h, discard the medium, add fresh culture medium (approximately 4 mL) and maintain the culture in the incubator (at 37 °C with a 5 % CO_2).
23. Replenish flasks with fresh medium (approximately 4 mL) every alternate day.
24. Monitor the growth of cells under an inverted phase-contrast microscope (Fig. 2a).
25. Repeat **steps 23–24**, until cells achieve 70–80 % confluency.
26. Discard medium after cells are confluent and wash cells attached to flasks twice with 1× PBS.
27. Add 1 mL Trypsin–EDTA into flasks and incubate for 5 min in a humidified incubator maintained at 37 °C with a 5 % CO_2 supply.
28. After 5 min, check that cells in the flask are dislodged.
29. Add 1 mL DMEM/F-12 supplemented with 10 % (v/v) FCS to cells to neutralize the Trypsin–EDTA, once cells are detached from the flask.
30. Transfer the cell suspension to a 15 mL tube and centrifuge at 200 × *g* for 5 min.
31. Discard the supernatant and resuspend the cell pellet in culture medium.
32. Cells can then be characterized, cultured, or cryopreserved.

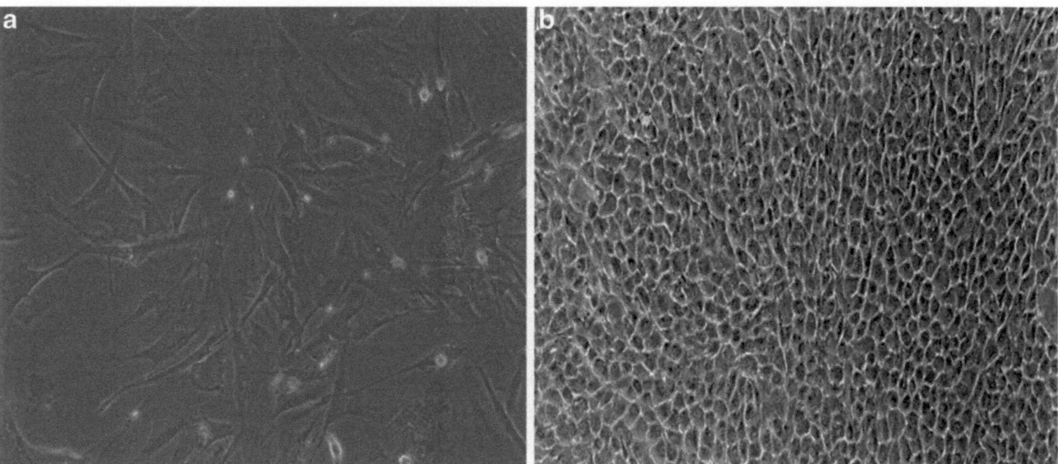

Fig. 2 (**a**) Phase-contrast micrograph of primary culture of limbal stromal cells (**a**) (100×) and corneal endothelial cells (**b**) (100×)

3.2 Isolation and Culture of Corneal Endothelial Cells

1. Wash corneas three times with M199 medium containing 50 μg/mL gentamycin and 1:100 diluted antibiotic/antimycotic solution (*see* **Note 5**).
2. After washing, place the corneas endothelial side up in a petri-dish.
3. Excise Descemet's membrane containing the intact endothelium in small strips with fine forceps (*see* **Note 6**).
4. Wash endothelium strips three times with M199.
5. After washing, incubate endothelium strips in culture medium overnight. This helps stabilize endothelial cells before culture.
6. After 24 h, centrifuge endothelium strips at 300 ×*g* for 5 min.
7. Discard the supernatant and transfer the endothelial strips to a culture dish.
8. Add 2–3 mL 0.02 % (w/v) EDTA solution to the culture dish and incubate at 37 °C for 1 h to loosen cell–cell junctions.
9. After the 1 h incubation period, isolate endothelial cells from Descemet's membrane by forcing the tissue and medium multiple times through the narrow opening of a flame-polished glass pipette. Pellet endothelial cells by centrifuging medium at 400 ×*g* for 5 min.
10. Discard the supernatant and resuspend the cell pellet in 1 mL culture medium.
11. Transfer the cell suspension to the well of a 12-well tissue culture plate, pre-coated with undiluted FNC coating mix (contains fibronectin, collagen, and albumin).
12. Incubate cultures in a humidified incubator maintained at 37 °C with a 5 % CO_2 supply.

13. Replace the medium every other day and monitor cell growth using a phase-contrast microscope (Fig. 2b).
14. Once cells are 70–80 % confluent, they should be sub-cultured in 1:2 ratios. A 0.02–0.05 % Trypsin–EDTA solution should be used to detach cells from culture flasks.
15. Cells can be characterized after cultivation.

4 Notes

1. Donor corneas from an eye bank that have been preserved for less than 48 h are preferred. Collect tissues only if they have passed the routine serological testing carried out in the eye bank. Approval from appropriate regulatory agencies and informed consent from donors are mandatory before collecting the tissue biopsy specimen.
2. Use a phase-contrast microscope to observe stromal tissue to ensure that sclera, transparent cornea and conjunctiva have been completely removed. This is to avoid contamination of conjunctival or scleral stromal cells.
3. Limbal segments must remain in culture medium whilst they are sectioned into explants to avoid drying out the sample. Cut limbal segments as small as possible, to enable efficient enzymatic digestion.
4. If the limbal tissue is not digested completely, increase the incubation time period in collagenase solution.
5. Exclude corneas that are obtained more than 24 h between time of death and preservation, corneas from donors with diabetes, glaucoma, sepsis, ocular infections, or corneas from donors who previously received large doses of chemotherapeutic agents. This will eliminate low endothelial cell densities.
6. Ensure that Descemet's membrane is excised with only the intact endothelium. Avoid isolation of stromal tissue together with Descemet's membrane.

References

1. Secker GA, Daniels JT (2009) Limbal epithelial stem cells of the cornea. *StemBook [internet]*, ed. The Stem Cell Research Community, StemBook, doi:10.3824/stembook.1.48.1
2. Tseng SC (1989) Concept and application of limbal stem cells. Eye 3:141–157
3. Kinoshita S, Friend J, Thoft RA (1981) Sex chromatin of donor corneal epithelium in rabbits. Invest Ophthalmol Vis Sci 21:434–441
4. Figueira EC, Di Girolamo N, Coroneo MT, Wakefield D (2007) The phenotype of limbal epithelial stem cells. Invest Ophthalmol Vis Sci 48:144–156
5. De Paiva CS, Pflugfelder SC, Li DQ (2006) Cell size correlates with phenotype and proliferative capacity in human corneal epithelial cells. Stem Cells 24:368–375
6. Akinci MA, Turner H, Taveras M, Barash A, Wang Z, Reinach P et al (2009) Molecular

profiling of conjunctival epithelial side-population stem cells: atypical cell surface markers and sources of a slow-cycling phenotype. Invest Ophthalmol Vis Sci 50:4162–4172

7. Schlotzer-Schrehardt U, Kruse FE (2005) Identification and characterization of limbal stem cells. Exp Eye Res 81:247–264
8. Itskovitz-Eldor J, Schuldiner M, Karsenti D, Eden A, Yanuka O, Amit M et al (2000) Differentiation of human embryonic stem cells into embryoid bodies compromising the three embryonic germ layers. Mol Med 6:88–95
9. Schofield R, Lajtha LG (1983) Determination of the probability of self-renewal in haemopoietic stem cells: a puzzle. Blood Cells 9:467–483
10. Li L, Neaves WB (2006) Normal stem cells and cancer stem cells: the niche matters. Cancer Res 66:4553–4557
11. Gipson IK (1989) The epithelial basement membrane zone of the limbus. Eye 3: 132–140
12. Ljubimov AV, Burgeson RE, Butkowski RJ, Michael AF, Sun TT, Kenney MC (1995) Human corneal basement membrane heterogeneity: topographical differences in the expression of type IV collagen and laminin isoforms. Lab Invest 72:461–473
13. Li DQ, Tseng SC (1995) Three patterns of cytokine expression potentially involved in epithelial-fibroblast interactions of human ocular surface. J Cell Physiol 163:61–79
14. Dravida S, Pal R, Khanna A, Tipnis SP, Ravindran G, Khan F (2005) The transdifferentiation potential of limbal fibroblast-like cells. Brain Res Dev Brain Res 160:239–251
15. Du Y, Funderburgh ML, Mann MM, SundarRaj N, Funderburgh JL (2005) Multipotent stem cells in human corneal stroma. Stem Cells 23:1266–1275
16. Polisetty N, Fatima A, Madhira SL, Sangwan VS, Vemuganti GK (2008) Mesenchymal cells from limbal stroma of human eye. Mol Vis 14:431–442
17. Criscimanna A, Zito G, Taddeo A, Richiusa P, Pitrone M, Morreale D et al (2011) *In vitro* generation of pancreatic endocrine cells from human adult fibroblast-like limbal stem cells. Cell Transplant 21:73–90
18. Nishida T, Nakamura M, Konma T, Ofuji K, Nagano K, Tanaka T et al (1997) Neurotrophic keratopathy–studies on substance P and the clinical significance of corneal sensation. Nippon Ganka Gakkai Zasshi 101:948–974
19. Bourne WM, Nelson LR, Hodge DO (1997) Central corneal endothelial cell changes over a ten-year period. Invest Ophthalmol Vis Sci 38:779–782
20. Williams KK, Noe RL, Grossniklaus HE, Drews-Botsch C, Edelhauser HF (1992) Correlation of histologic corneal endothelial cell counts with specular microscopic cell density. Arch Ophthalmol 110:1146–1149
21. Blake DA, Yu H, Young DL, Caldwell DR (1997) Matrix stimulates the proliferation of human corneal endothelial cells in culture. Invest Ophthalmol Vis Sci 38:1119–1129
22. Yue BY, Sugar J, Gilboy JE, Elvart JL (1989) Growth of human corneal endothelial cells in culture. Invest Ophthalmol Vis Sci 30:248–253
23. Miyata K, Drake J, Osakabe Y, Hosokawa Y, Hwang D, Soya K et al (2001) Effect of donor age on morphologic variation of cultured human corneal endothelial cells. Cornea 20:59–63
24. Amano S (2003) Transplantation of cultured human corneal endothelial cells. Cornea 22:S66–S74
25. Pistsov MY, Sadovnikova E, Danilov SM (1988) Human corneal endothelial cells: isolation, characterization and long-term cultivation. Exp Eye Res 47:403–414
26. Zhu C, Joyce NC (2004) Proliferative response of corneal endothelial cells from young and older donors. Invest Ophthalmol Vis Sci 45:1743–1751
27. Engelmann K, Friedl P (1989) Optimization of culture conditions for human corneal endothelial cells. In Vitro Cell Dev Biol 25:1065–1072
28. Dawson DG, Kramer TR, Grossniklaus HE, Waring GO, Edelhauser HF (2005) Histologic, ultrastructural, and immunofluorescent evaluation of human laser-assisted in situ keratomileusis corneal wounds. Arch Ophthalmol 123:741–756
29. Ishino Y, Sano Y, Nakamura T, Connon CJ, Rigby H, Fullwood NJ et al (2004) Amniotic membrane as a carrier for cultivated human corneal endothelial cell transplantation. Invest Ophthalmol Vis Sci 45:800–806
30. Choi JS, Williams JK, Greven M, Walter KA, Laber PW, Khang G et al (2010) Bioengineering endothelialized neo-corneas using donor-derived corneal endothelial cells and decellularized corneal stroma. Biomaterials 31:6738–6745
31. Chen KH, Azar D, Joyce NC (2001) Transplant of adult human corneal endothelium ex vivo: a morphologic study. Cornea 20:731–737
32. Joyce NC, Zhu CC (2004) Human corneal endothelial cell proliferation: potential for use in regenerative medicine. Cornea 23:S8–S19

Part IV

Development of the Artificial Cornea

Chapter 9

The Formation of a Tissue-Engineered Cornea Using Plastically Compressed Collagen Scaffolds and Limbal Stem Cells

Shengli Mi and Che J. Connon

Abstract

Collagen has excellent biocompatibility, is biodegradable, and possesses low immunogenicity. Therefore, this protein is a very suitable substrate for the formation of a corneal scaffold for therapeutic use. The highly hydrated nature of conventional collagen gels, however, results in a gel that is structurally weak and difficult to manipulate. In this chapter, we describe a novel method to cultivate limbal epithelial cells (LEC) on a compressed collagen scaffold. The compressed collagen scaffold can be rapidly constructed using a cell-independent process, which produces dense and mechanically strong collagen constructs with controllable microscale features.

We embedded corneal keratocytes in a collagen gel, which we subsequently compressed and coated with laminin. The resulting construct supported the physiological morphology and stratification of LEC. The expression of a specific marker for differentiated LEC, cytokeratin 3 (CK3), and a marker for undifferentiated LEC, cytokeratin 14 (CK14), were similar in LEC expanded on both the compressed collagen construct and the leading conventional scaffold, denuded amniotic membrane (AM). We therefore demonstrate that a laminin-coated, compressed collagen gel containing keratocytes can support LEC expansion, stratification, and differentiation to a degree that is comparable to denuded AM. Our novel compressed collagen/keratocyte construct has potential for use as a tissue-engineered artificial cornea.

Key words Compressed collagen gel, Tissue-engineered cornea, Limbal epithelial cells, Limbal keratocytes, Compressed collagen/keratocyte construct, Denuded amniotic membrane

1 Introduction

The concept of the artificial corneal substitute was introduced in 1789 by Guillaume Pellier de Quengsy as a keratoprosthesis (Kpro) with a porous prosthetic skirt [1] (*see* Chapter 2). This design proved to be fundamental to artificial cornea research and is currently the model used for the Chirila Kpro, AlphaCor™ [2–5], one of the most clinically successful corneal substitutes. The biocompatibility of the porous skirt material is key to the success of Kpro

Bernice Wright and Che J. Connon (eds.), *Corneal Regenerative Medicine: Methods and Protocols*, Methods in Molecular Biology, vol. 1014, DOI 10.1007/978-1-62703-432-6_9, © Springer Science+Business Media New York 2013

in repairing the cornea [2, 3, 5]. Microporous materials which allow corneal tissue ingrowth including poly(2-hydroxyethyl methacrylate) (PHEMA) [2–6], porous nano-hydroxyapatite/poly (vinyl alcohol) [7–10], and collagen hydrogels [11–23] have ensured the biocompatibility of Kpro.

Collagen hydrogels, in particular, have replaced less biocompatible materials used for the construction of biointegrable Kpro [11–23], and are currently under development for clinical application [14, 23]. Collagen type I, the most abundant stromal protein in the cornea [24], is biocompatible and biodegradable and possesses low immunogenicity. Collagen as an extracellular matrix (ECM) mimetic is, therefore, appropriate for use in the formation of a tissue-engineered corneal scaffold.

Conventional collagen gels are inherently weak and difficult to manipulate as fibers of this protein are dispersed in a highly hydrated form [25]. Chemical cross-linking is a common strategy to increase the mechanical strength of collagen hydrogels. Collagen gels cross-linked with 1-ethyl-3-(3-dimethyl aminopropyl) carbodiimide (EDC) and *N*-hydroxysuccinimide (NHS) [21–23] and *N*-isopropylacrylamide, acrylic acid, and acryloxysuccinimide (*see* Chapter 10) [16] were previously demonstrated to promote regeneration of corneal cells, nerves, and the tear film. EDC and NHS cross-linked collagen was also tested on humans in a Phase 1 clinical trial with promising results [23].

A more practical and less labor-intensive method for enhancing the mechanical strength of collagen gels is through plastic compression, which eliminates the majority of the water content of this gel. In this chapter, we describe a protocol for the plastic compression of collagen hydrogels intended for application as an artificial cornea. By varying collagen concentration and compression times, we constructed a stronger collagen-based substrate. We performed controlled, unconfined plastic compression to rapidly produce dense, pliable, mechanically strong collagen scaffolds with controllable microscale features [26]. In order to mimic the normal corneal stroma and produce stratified layers of LEC on compressed collagen gels, corneal keratocytes were embedded in the collagen scaffold before compression and the compressed collagen gel was coated with laminin. This engineered structure provides an improved substrate capable of supporting the expansion and stratification of LEC, applicable to both corneal repair and corneal replacement.

2 Materials

Prepare all solutions in a class II biological safety cabinet using sterile (*see* **Note 1**), ultrapure water (prepared by purifying deionized water to attain a sensitivity of 18 MΩ cm at 25 °C) and analytical grade reagents. Store prepared solutions at room temperature unless indicated otherwise.

2.1 Human and Bovine Cornea

Normal bovine eyes can be obtained from a local abattoir (*see* **Note 2**). Human corneoscleral rims (*see* **Note 3**) which are available after transplantation of the corneal button can be obtained from a local hospital. Human corneas are also available from eye banks. UK-based eye banks include the Bristol Eye Bank and the Manchester Eye bank, and those are also located at the Moorfields Eye Hospital (London, UK) and the Queen Victoria Hospital (East Grinstead, West Sussex, UK).

2.2 Cell Detachment from the Cornea

1. Class II biological safety cabinet (Triple Red Laboratory Technology: Buckinghamshire, UK).
2. Cell culture CO_2, 37 °C incubator containing a High-Efficiency Particulate Air (HEPA) filter (Triple Red).
3. 0.02 % (w/v) Collagenase: 2 mg collagenase (GIBCO (Invitrogen): Paisley, UK) dissolved in 10 mL Dulbecco's minimal essential medium (DMEM) and Ham's F12 medium (DMEM/F12, 1:1) (GIBCO).
4. 0.25 % (w/v) Trypsin solution containing 0.02 % (w/v) ethylenediaminetetraacetic acid (EDTA).
5. Limbal epithelial cell and keratocyte culture medium (*see* Subheadings 2.3 and 2.4).
6. Tissue dissection kit containing fine forceps, straight scissors (length: 11.5 cm) with very fine points, sterile disposal razor blades, and scalpels.
7. Sterile 85 mm diameter Petri dishes.
8. Dissection microscope. We use a Leica S68 dissecting microscope (Leica Microsystems Ltd.: Milton Keynes, UK) that is custom-fitted into our biological safety cabinet.
9. 21 G needles.
10. 2 and 1 mL syringes.
11. Gentle MACS™ dissociator (Miltenyi Biotec: Surrey, UK).
12. Cell strainers.
13. Hemocytometer.
14. Automated cell counter (Bio-Rad (Hercules, CA, USA), Invitrogen, Millipore (Watford, UK)).
15. Trypan blue dye solution: (0.4 % (w/v) Trypan blue dye in phosphate-buffered saline (PBS)).
16. Number 0 or number 1 glass coverslips.
17. Inverted phase contrast microscope.
18. Water bath set to 37 °C.

2.3 Keratocyte Culture Medium

Basal medium containing DMEM supplemented with B27 (Invitrogen), 10 % (w/v) fetal bovine serum (FBS), 40 ng/mL basic fibroblast growth factor, 100 U/mL penicillin, 100 μg/mL streptomycin, and 250 ng/mL amphotericin B (*see* **Note 4**).

2.4 Limbal Epithelial Cell Culture Medium

Basal medium containing DMEM/F12 (1:1) (GIBCO) supplemented with 10 % (w/v) FBS, 0.5 % (v/v) dimethylsulfoxide, 10 ng/mL epithelial growth factor (EGF), 5 mg/mL insulin, 100 IU/mL penicillin, 100 mg/mL streptomycin, and 250 ng/mL amphotericin B (*see* **Note 4**).

2.5 Collagen Gel Components

1. Sterile rat-tail type I collagen (2.10 mg/mL) (First link: Birmingham, UK).
2. 10× Concentrate Eagle's minimum essential medium (MEM: GIBCO).
3. 1 M Sodium hydroxide.
4. Laminin solution (50 μg/mL).
5. Rectangular gel molds (33 mm×13 mm×4 mm) (custom made in-house).
6. Sterile nylon gauze.
7. Whatman 185 MM chromatography paper.
8. 120 g Glass weight (dimensions: 85 mm×79 mm×5 mm) (custom made in-house).
9. 6-well Transwell culture plates.

2.6 Preparation of Denuded AM

1. Sterile 140 mm diameter Petri dishes.
2. Tissue dissection kit containing fine forceps, straight scissors (length: 11.5 cm) with very fine points, sterile disposal razor blades, and scalpels.
3. Sterile cell scrapers.
4. Sterile PBS.
5. 0.25 % (w/v) Trypsin containing 0.02 % (w/v) EDTA.
6. DMEM.

2.7 Keratocyte Viability

1. Live/dead fluorescent cell staining kit (Calbiochem).
2. 35 mm diameter Petri dishes.
3. PBS.
4. Fine forceps and straight, fine-tipped scissors.
5. Glass bottom dishes with number 0 or number 1 glass inserts (SLS, MatTek (Ashland, MA, USA), Greiner Bio-One).
6. Number 0 or number 1 glass coverslips.
7. Aluminum foil.
8. Confocal microscope (Leica Microsystems Ltd.).

2.8 Immunohistochemical Analysis

1. Optimal cutting temperature (OCT) tissue freezing medium.
2. Polystyrene molds for embedding tissue samples in OCT.
3. −80 °C freezer.
4. Polysine adhesion glass slides (Fisher Scientific).
5. Cryostat microtome (Leica Microsystems Ltd, Zeiss (Hertfordshire, UK)).
6. 1 % (w/v) Bovine serum albumin (BSA): 0.02 g BSA dissolved in 1 mL PBS solution.
7. 0.02 % (v/v) Tween-20 (Sigma): 10 μL Tween-20 mixed with 490 mL PBS solution (GIBCO).
8. Anti-CK3 antibody (1:50 dilution) (Chemicon) and anti-CK14 antibody (1:100 dilution) (Progen Scientific).
9. Humidified container.
10. Fluorescein isothiocyanate (FITC)-labeled secondary anti-mouse and anti-guinea pig antibodies (1:50 dilution) (Progen Scientific, Cell Signalling Technology, R and D systems), GE Healthcare, Santa Cruz (Autogen Bioclear) and Jackson Laboratories (Stratech). Antibodies may be conjugated to other fluorescent tags including rhodamine:tetramethyl rhodamine isothiocyanate and fluorescent proteins such as green fluorescent protein, allophycocyanin, phycocyanin, phycoerythrin, and phycoerythrocyanin.
11. Mounting medium with PI or 4′,6-diamidino-2-phenylindole (DAPI) (Vector Laboratories Ltd.).
12. Number 0 or number 1 glass coverslips.
13. Aluminum foil.
14. Fluorescence microscope fitted with a camera (Leica Microsystems Ltd, Zeiss).

3 Methods

Perform all procedures in a class II biological safety cabinet, at room temperature unless otherwise specified.

3.1 Isolation of Limbal Epithelial Cells

1. Rinse dissected whole corneoscleral tissues three times with DMEM, before carefully removing the central cornea, excess sclera, iris, corneal endothelium, conjunctiva, and Tenon's capsule with a razor blade and a scalpel.
2. Cut the limbal rim that remains into small pieces (explants: approximately 2.5 mm^2).
3. Digest limbal explants overnight in 0.02 % (w/v) collagenase to loosen the epithelium.

4. Carefully peel the epithelium away from the limbal stroma with fine forceps. This step should be performed using a dissection microscope inside the safety cabinet.
5. Incubate limbal epithelial sheets (at 37 °C, 5 % CO_2, 95 % humidity) in 0.25 % (w/v) trypsin and 0.02 % (w/v) EDTA (cell digest solution) for 5 min.
6. Neutralize the cell digest solution with an equal volume of limbal epithelial cell culture medium.
7. Aspirate digested limbal epithelium sheets through a 21 G needle to obtain a single-cell suspension.
8. Centrifuge the cell suspension for 10 min at 453 × *g* (1,500 rpm). Discard the supernatant and wash cells with limbal epithelial cell culture medium by resuspending the cell pellet in this medium, before a final centrifugation (10 min, 453 × *g* (1,500 rpm)) step.
9. Use a manual hemocytometer or an automated cell counter (*see* **Note 5**) to assess cell number and viability. Mix equal volumes of 0.4 % (w/v) Trypan blue stain (*see* **Note 6**) and a well-mixed cell suspension. Pipette the Trypan blue/cell mix (10 μL) at the edge of the coverslip and allow this to run under the coverslip. View the hemocytometer grid under an inverted phase contrast microscope. Count live (unstained) cells and dead (blue stained) cells using a cell counter. Refer to the manufacturer's instructions for the layout/dimensions of the hemocytometer grid.
10. Culture cells in limbal epithelial culture medium (at 37 °C, 5 % CO_2, 95 % humidity) in preparation for expansion on compressed collagen scaffolds or denuded AM.
11. Tissue that is not used for experiments should be disposed of by incineration.

3.2 Isolation of Limbal Keratocytes

1. Digest limbal explants overnight with 0.02 % (w/v) collagenase in the cell culture incubator (at 37 °C, 5 % CO_2, 95 % humidity).
2. Remove limbal epithelial cell sheets, and treat limbal stroma remaining with 0.25 % (w/v) trypsin and 0.02 % (w/v) EDTA (cell digest solution) at 37 °C (in a water bath) for 15 min.
3. Neutralize the cell digest solution with an equal volume of keratocyte culture medium.
4. Release keratocytes from digested stroma through gentle homogenization using the gentle MACS™ dissociator or an equivalent instrument. Separate cells from homogenized tissue into a 50 mL falcon tube using a cell strainer.
5. Wash cells as described in Subheading 3.1, **step 8**.

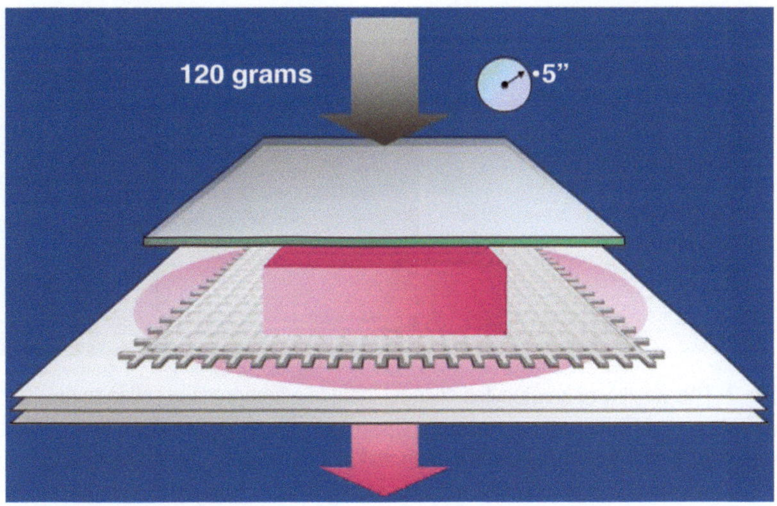

Fig. 1 Plastic compression of a collagen gel

6. Count cells and assess cell viability as described in Subheading 3.1, **step 9**.
7. Culture cells (at 37 °C, 5 % CO_2, 95 % humidity) in keratocyte culture medium (*see* **Note 7**) in preparation for encapsulation in collagen gels.

3.3 Preparation of Compressed Collagen Scaffold

1. Gently mix together 4 mL sterile rat-tail type I collagen, 0.5 mL 10× concentrate Eagle's MEM, 0.5 mL keratocyte (2.0×10^5 cells/mL) suspension, and 0.5 mL 1 M sodium hydroxide (*see* **Note 8**).
2. Cast the gel solution (*see* **Note 9**) into gel molds and allow the gel to set in the cell culture incubator (at 37 °C, 5 % CO_2, 95 % humidity) for 30 min.
3. Compact gels through compression and dehydration (Fig. 1). Place a layer of nylon mesh onto a double layer of absorbent chromatography paper. Place the collagen gel on the nylon mesh, cover the gel with a second nylon mesh, and load the assembly with a 120 g glass weight for 5 min at room temperature. This will result in the formation of a flat collagen sheet protected between two nylon meshes.
4. Transfer the compressed collagen/keratocyte construct to the inserts of 6-well Transwells.
5. Coat the compressed collagen scaffolds with laminin, by incubating constructs with the protein solution at 37 °C (in the cell culture incubator) for 2 h.
6. Wash gels three times with PBS in preparation for LEC expansion.

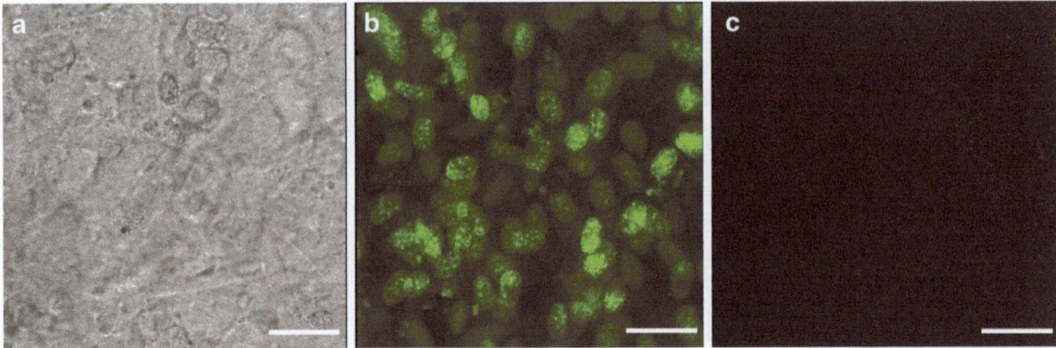

Fig. 2 **The** viability of keratocytes encapsulated in a compressed collagen gel. Keratocytes (**a**) remain viable (cyto-dye-stained) (**b**) within a compressed collagen gel after a 7-day culture period. PI-stained, dead keratocytes are not visible (**c**). Cyto-dye fluorescence is detected at 518 nm after excitation at 488 nm and PI fluorescence is detected at 615 nm after excitation at 488 nm with an argon laser. Images of a single middle layer from z-stacks are shown. Images represent three different experiments from three different corneoscleral rims. Scale bar: 100 μm

3.4 Analysis of Keratocyte Viability

It is necessary to validate the viability of keratocytes encapsulated in compressed collagen gel scaffolds. This can be achieved using a live/dead fluorescent cell staining procedure and confocal microscopy to image encapsulated cells. We stained cells after a 7-day culture period to achieve a representative indication of the viability of keratocytes within gels (Fig. 2). The live/dead staining kit includes Cyto-dye (Ex. max.: 488 nm; Em. max.: 518 nm), a cell-permeable green fluorescent dye to stain live cells, and dead cells are stained by PI (Ex. max.: 488 nm; Em. max.: 615 nm), a red fluorescent dye.

1. Wash a piece of compressed collagen gel (approximately 5 mm^2) containing keratocytes with PBS in a 35 mm diameter petri dish.
2. Dilute 5 μL Cyto-dye in 1 mL PBS and pipette 300–500 μL diluted Cyto-dye onto the piece of gel construct.
3. Wash the stained scaffold with PBS to remove excess Cyto-dye stain.
4. Dilute 5 μL PI in 1 mL PBS and pipette 300–500 μL diluted PI onto the piece of gel construct.
5. Wash the stained scaffold with PBS to remove excess PI stain.
6. Spread the stained gel scaffold on the glass insert of a glass-bottom plate and cover the gel with a glass coverslip. Cover the petri dish with foil to prevent bleaching of the fluorescence signal.
7. Image the stained gel using a confocal microscope set at ×63 magnification. Obtain a series of images in the *z*-dimension at 0.2 μm intervals throughout the gel.

3.5 Preparation of Denuded AM

1. Collect the placenta and fetal membranes (AM) of healthy women, postpartum.
2. Separate AM and chorionic tissues manually (using a tissue dissection kit) and wash the AM with chilled, sterile PBS to remove blood.
3. Treat the AM with 0.25 % (w/v) trypsin at 37 °C (in the cell culture incubator) for 30 min.
4. Remove AM epithelial cells with a cell scraper.
5. Transfer epithelial cell-free AM to the inserts of 6-well Transwells with the basement membrane surface facing up.

3.6 Expansion of Limbal Epithelial Cells on Collagen Gels and Denuded AM

1. Seed isolated LEC (10^6 cells/mL) onto the laminin-coated, compressed collagen/keratocyte construct and onto denuded AM.
2. Expose expanded LEC to air (airlift) after 10 days by lowering the medium so that it is level with the collagen substrate. Airlift cells for 4 days.

3.7 Immunohistochemical Analysis

To demonstrate that the tissue-engineered collagen scaffold is comparable to AM for the expansion of LEC, the phenotype of LEC grown on collagen constructs should be validated. The expression of CK3, a specific marker for corneal epithelial cells, and CK14, a marker of undifferentiated, basal epithelial cells, can be examined to indicate cell phenotype (Fig. 3).

1. Embed collagen gel constructs in OCT and freeze those at −80 °C.
2. Cryosection (*see* **Note 10**) the OCT-embedded gel and mount tissue slices on polysine-coated glass slides.
3. Block tissue sections with 2 % (w/v) BSA for 30 min at room temperature.
4. Incubate tissue sections with primary anti-CK3 and anti-CK14 antibodies overnight at 4 °C in a humidified container (*see* **Note 11**).
5. Wash mounted tissue with 0.02 % (v/v) Tween-20 for 5 min. Repeat this wash step twice.
6. Incubate tissue sections with secondary anti-mouse and anti-guinea pig FITC-labeled antibodies for 1 h at room temperature in a darkened, humidified container (*see* **Note 12**).
7. Wash mounted tissue with 0.02 % (v/v) Tween-20 for 5 min. Repeat this wash step twice.
8. Co-stain tissue sections with PI or DAPI mounting medium (Vectorsheild®), cover with glass coverslips, and image cells using a fluorescence microscope.

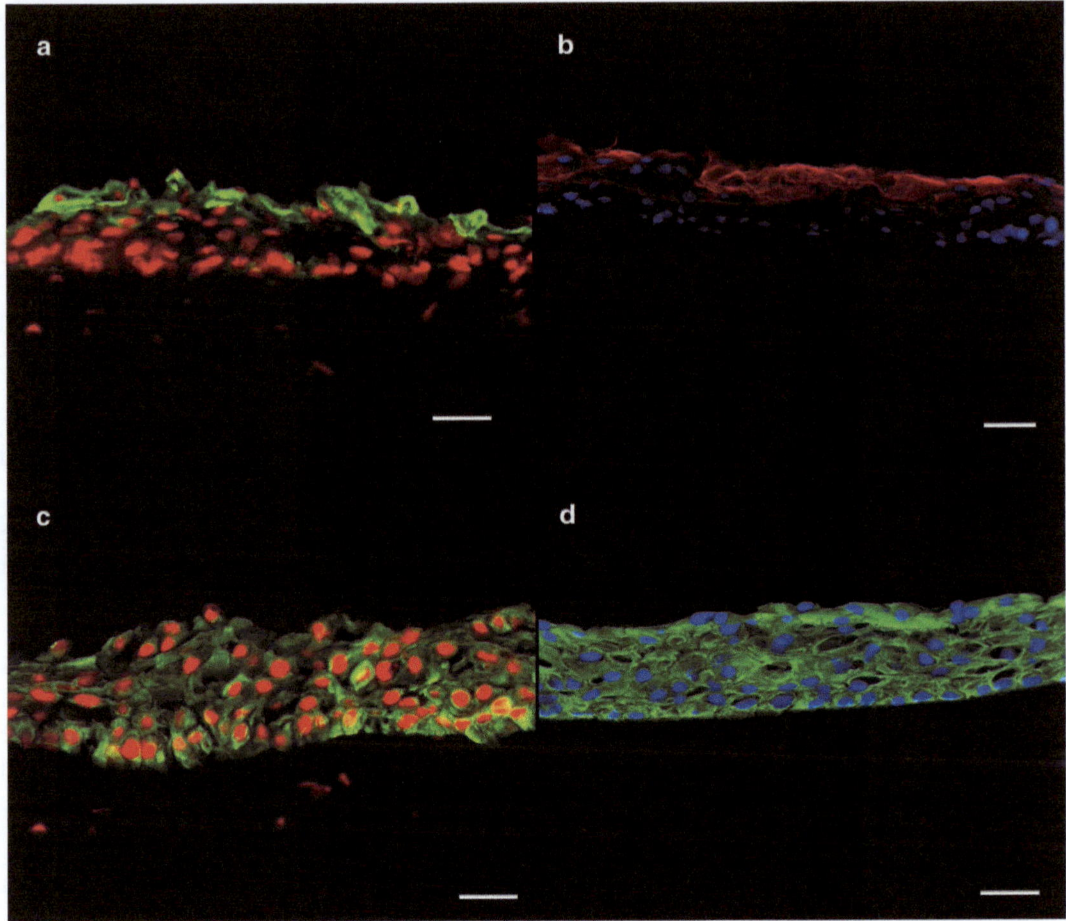

Fig. 3 CK3 and CK14 expression in LEC expanded on a compressed collagen/keratocyte construct and denuded AM. CK3 expression in LECs (*green*) expanded on a laminin-coated compressed collagen gel embedded with keratocytes (**a**) and on denuded AM (CK3 *red*) (**b**) is similar. CK14 (*green*) is also expressed to similar levels on a compressed collagen/keratocyte construct (**c**) and denuded AM (**d**). Cell nuclei are stained with PI or DAPI. Rhodamine-labelled CK3 fluorescence and PI are detected at 615 nm after excitation at 595–605 nm, FITC-labelled CK14 fluorescence is detected at 518 nm after excitation at 488 nm, and DAPI fluorescence is detected at 456–460 nm after excitation at 360 nm with a laser. Images represent three different experiments from three different corneoscleral rims. Scale bar: 50 μm

4 Notes

1. Most apparatus are sterilized upon purchase, but some reagents and equipment which are not can be sterilized by subjecting them to high-pressure saturated steam at 121 °C (autoclaving).
2. Bovine eyes should be collected within 2 h of death and transported to the laboratory on ice. Eyes should ideally be used immediately but they can be used the following day if they are stored on ice at 4 °C.

3. Considerable difficulty has been encountered with the culture of human LEC using methods which do not involve AM. A variety of media including EpiLife (Invitrogen), CnT20 (CellNTec: Buckinghamshire, UK), and Keratinocyte Serum-Free Medium (KSFM: Invitrogen) have been used in the successful culture of these cells.
4. Culture LEC and keratocytes at 37 °C under 5 % CO_2 and 95 % humidity and replenish cultures with fresh medium every 2 days.
5. Typically we use a manual hemocytometer to count cells. The cell concentration per mL is calculated using the following equation: average number of cells in one large square × dilution factor × 10^4.
6. Trypan blue is a vital dye that is excluded from live cells and incorporated into dead cells. Another well-established cell viability assay is the 12 mM 3-(4,5-dimethylthiazol-2-yl)-2,5-diphenyltetrazolium bromide (MTT) assay. The MTT assay measures the mitochondrial activity of cells. The MTT reagent is added to a 100 μL suspension of cells and this is incubated for 2 h at 37 °C. Cells are then lysed using dimethylsulfoxide, incubated for a further 10 min at 37 °C, and mixed and the absorbance of the reaction product (formazan) is measured at 540 nm using a spectrophotometer. A standard curve derived from known cell numbers is used to calculate the number of cells in test samples. We use the Trypan blue assay, as the MTT assay requires greater number of cells which are sometimes in limited supply.
7. Stromal keratocytes are transformed into fibroblasts when removed from in vivo tissues. Culture conditions containing minimal levels of serum or specialized medium (e.g., KSFM, CnT20) can reduce this transformation.
8. To make the collagen gel with encapsulated keratocytes, first dispense 4 mL sterile rat-tail type I collagen in a centrifuge tube, then add 0.5 mL 10× concentrate Eagle's MEM, and mix the solution gently. Next add 0.5 mL 1 M sodium hydroxide to adjust the pH, add 0.5 mL keratocytes to the suspension, and gently mix.
9. Mix the collagen gel solution gently and place on ice for 30 min before casting to prevent gelling whilst allowing the dispersion of any bubbles within the solution.
10. Limit the thickness of tissue sections to approximately 5 μm.
11. Ensure that antibody solutions totally cover tissue sections on the slides.
12. Cover the humidified container with foil to prevent bleaching of the fluorescence signal from secondary antibodies.

Acknowledgment

This study is financially supported by the Biotechnology and Biological Sciences Research Council (BBSRC).

References

1. Chirila TV, Hicks CR (1999) The origins of the artificial cornea: Pellier de Quengsy and his contribution to the modern concept of keratoprosthesis. Gesnerus 56:96–106
2. Vijayasekaran S, Fitton JH, Hicks CR, Chirila TV, Crawford GJ, Constable IJ (1998) Cell viability and inflammatory response in hydrogel sponges implanted in the rabbit cornea. Biomaterials 19:2255–2267
3. Sandeman SR, Faragher RGA, Aleen MCA, Liu C, Lloyd AW (2000) Novel materials to enhance keratoprosthesis integration. Br J Ophthalmol 84:640–644
4. Hicks CR, Crawford GJ, Lou X, Tan DT, Snibson GR, Sutton G et al (2003) Corneal replacement using a synthetic hydrogel cornea, AlphaCorTM: device, preliminary outcomes and complications. Eye 17:385–392
5. Chirila TV (2001) An overview of the development of artificial corneas with porous skirts and the use of PHEMA for such an application. Biomaterials 22:3311–3317
6. Crawford GJ, Chirila TV, Vijayasekaran S, Dalton PD, Constable IJ (1996) Preliminary evaluation of a hydrogel core-and-skirt keratoprosthesis in the rabbit cornea. J Refract Surg 12:525–529
7. Fenglan X, Yubao L, Xiaoming Y, Hongbing L, Li Z (2007) Preparation and in vivo investigation of artificial cornea made of nano-hydroxyapatite/poly(vinyl alcohol) hydrogel composite. J Mater Sci Mater Med 18:635–640
8. Trinkaus-Randall V, Capecchi J, Newton A, Vadasz A, Leibowitz H, Franzblau C (1988) Development of a biopolymeric keratoprosthetic material. Evaluation in vitro and in vivo. Invest Ophthalmol Vis Sci 29:393–400
9. Tsuk AG, Trinkaus-Randall V, Leibowitz HM (1997) Advances in polyvinyl alcohol hydrogel keratoprostheses: protection against ultraviolet light and fabrication by a molding process. J Biomed Mater Res 34:299–304
10. Xu F, Li Y, Deng Y, Xiong J (2008) Porous nano-hydroxyapatite/poly(vinyl alcohol) composite hydrogel as artificial cornea fringe: characterisation and evaluation in vitro. J Biomater Sci Polym Ed 19:431–439
11. Li F, Griffith M, Li Z, Tanodekaew S, Sheardown H, Hakim M et al (2005) Recruitment of multiple cell lines by collagen-synthetic copolymer matrices in corneal regeneration. Biomaterials 26:3093–3104
12. Hackett JM, Lagali N, Merrett K, Edelhauser H, Sun Y, Gan J et al (2011) Biosynthetic corneal implants for replacement of pathologic corneal tissue: performance in a controlled rabbit alkali burn model. Invest Ophthalmol Vis Sci 52:651–657
13. Levis HJ, Brown RA, Daniels JT (2010) Plastic compressed collagen as a biomimetic substrate for human limbal epithelial cell culture. Biomaterials 31:7726–7737
14. Builles N, Janin-Manificat H, Malbouyres M, Justin V, Rovere MR, Pellegrini G et al (2010) Use of magnetically oriented orthogonal collagen scaffolds for hemi-corneal reconstruction and regeneration. Biomaterials 31:8313–8322
15. Tanaka Y, Kubota A, Matsusaki M, Duncan T, Hatakeyama Y, Fukuyama K et al (2011) Anisotropic mechanical properties of collagen hydrogels induced by uniaxial-flow for ocular applications. J Biomater Sci Polym Ed 22:1427–1442
16. Li F, Carlsson D, Lohmann C, Suuronen E, Vascotto S, Kobuch K et al (2003) Cellular and nerve regeneration within a biosynthetic extracellular matrix for corneal transplantation. Proc Natl Acad Sci U S A 100:15346–15351
17. Doillon CJ, Watsky MA, Hakim M, Wang J, Munger R, Laycock N et al (2003) A collagen-based scaffold for a tissue engineered human cornea: physical and physiological properties. Int J Artif Organs 26:764–773
18. Mi S, Chen B, Wright B, Connon CJ (2010) Plastic compression of a collagen gel forms a much improved scaffold for ocular surface tissue engineering over conventional collagen gels. J Biomed Mater Res A 95:447–453
19. Mi S, Khutoryanskiy VV, Jones RR, Zhu X, Hamley IW, Connon CJ (2011) Photochemical cross-linking of plastically compressed collagen gel produces an optimal scaffold for corneal tissue engineering. J Biomed Mater Res A 99:1–8

20. Mi S, Chen B, Wright B, Connon CJ (2010) Ex vivo construction of an artificial ocular surface by combination of corneal limbal epithelial cells and a compressed collagen scaffold containing keratocytes. Tissue Eng Part A 16:2091–2100
21. Liu W, Merrett K, Griffith M, Fagerholm P, Dravida S, Heyne B et al (2008) Recombinant human collagen for tissue engineered corneal substitutes. Biomaterials 29:1147–1158
22. Liu W, Deng C, McLaughlin CR, Fagerholm P, Lagali NS, Heyne B et al (2009) Collagen-phosphorylcholine interpenetrating network hydrogels as corneal substitutes. Biomaterials 30:1551–1559
23. Fagerholm P, Lagali NS, Merrett K, Jackson WB, Munger R, Liu Y et al (2010) A biosynthetic alternative to human donor tissue for inducing corneal regeneration: 24-month follow-up of a phase 1 clinical study. Sci Transl Med 2:46–61
24. Ihanamaki T, Pelliniemi LJ, Vuorio E (2004) Collagens and collagenrelated matrix components in the human and mouse eye. Prog Retin Eye Res 23:403–434
25. Chandran PL, Barocas VH (2004) Microstructural mechanics of collagen gels in confined compression: poroelasticity, viscoelasticity and collapse. J Biomech Eng 126:152–166
26. Brown RA, Wiseman M, Chuo CB, Cheema U, Nazhat SN (2005) Ultrarapid engineering of biomimetic materials and tissues: fabrication of nano- and microstructures by plastic compression. Adv Funct Mater 15:1762–1770

Chapter 10

Fabrication of a Human Recombinant Collagen-Based Corneal Substitute Using Carbodiimide Chemistry

Mohammad Mirazul Islam, May Griffith, and Kimberley Merrett

Abstract

Human recombinant collagen can be cross-linked with a variety of chemical cross-linking agents. Cross-linking methods can be tuned to confer collagen-based scaffolds with specific physical properties, improved antigenicity and thermal stability without impeding the ability of the material to integrate into the surrounding tissue and to promote regeneration. Here, we describe a method to cross-link human recombinant collagen using a water soluble carbodiimide. Carbodiimides are referred to as zero-length cross-linking agents as they are not incorporated into the final cross-link and thus pose minimal risk with respect to cytotoxicity. The resulting collagen-based scaffold possesses properties comparable to that of the human cornea and is thus suitable for use as a corneal substitute.

Key words Biosynthetic corneal implant, Collagen, Carbodiimide, Collagen cross-linking, Collagen-based scaffold

1 Introduction

Collagen is the predominant component of the corneal extracellular matrix (ECM), and therefore, it is the material of choice in the research and development of corneal substitutes (*see* Chapters 2 and 9). Collagen promotes cellular adhesion, and is biocompatible and weakly antigenic. Collagen is readily available from animal sources (e.g., porcine and bovine). Alternatively, highly purified recombinant human collagen can be produced using a yeast expression system or from bio-engineered plant lines (e.g., tobacco plants) [1–3]. Purified collagen, however, forms a weakly cross-linked thermogel lacking the sufficiently robust physical properties and resistance to biodegradation to make it an adequate material for corneal substitutes [4]. Hence, there is a requirement to use a synthetic cross-linking process that can confer purified collagen with suitable properties. It has been previously demonstrated that cross-linking collagen improves antigenicity [5], thermal stability [6], mechanical

Bernice Wright and Che J. Connon (eds.), *Corneal Regenerative Medicine: Methods and Protocols*, Methods in Molecular Biology, vol. 1014, DOI 10.1007/978-1-62703-432-6_10,

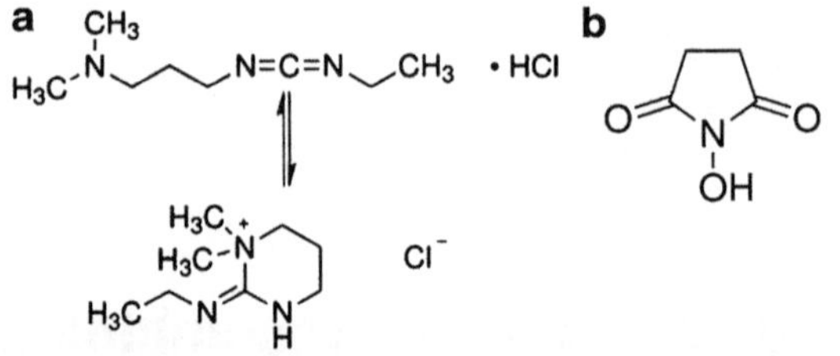

Fig. 1 (**a**) *N*-(3-Dimethylaminopropyl) *N′*-ethylcarbodiimide hydrochloride (EDC) (Molecular Weight: 191.70), (**b**) *N*-Hydroxysuccinimide (NHS) (Molecular Weight: 115.09)

strength [7], and the resistance of collagen to biodegradation [8] without impeding the ability of the material to integrate into the surrounding host tissue and to promote regeneration.

Numerous chemical cross-linking methods have been developed; each method employing different mechanisms. The choice of chemical cross-linking agent is very important in the development of materials intended for in vivo use. Chemical cross-linkers are potentially cytotoxic; for example, the commonly used cross-linker glutaraldehyde is known to be carcinogenic [9].

In this protocol we have chosen to use a water-soluble zero-length cross-linking agent, *N*-(3-dimethylaminopropyl) *N′*-ethylcarbodiimide hydrochloride (EDC or EDAC: Fig. 1a) to cross-link collagen to produce a robust, transparent material that can be cast with the appropriate dimensions for use as a corneal substitute. A zero-length reagent was chosen to cross-link collagen as no additional chemical structure is introduced between the conjugating groups thus minimizing concerns related to cytotoxicity. *N*-hydroxysuccinimide (NHS: Fig. 1b), or its water soluble analog sulfo-NHS is included in the EDC cross-linking protocol to improve cross-linking efficiency [10].

EDC reacts with the carboxylic acids of glutamic and aspartic acid residues on the collagen structure to form highly reactive *o*-acylisourea derivatives that are extremely short-lived. These active species react with the primary amines of lysine and/or hydroxyl-lysine residues to form a covalent amide bond [11]. Amide bond formation occurs with the highest yield at pH 4–6 [12]. When working with collagen, EDC mediated cross-linking effectively occurs between pH 4.5 and 5.5. We have investigated these reactions and produced a method for the cross-linking of collagen with EDC and NHS that is applicable as a corneal substitute.

2 Materials

Prepare all solutions using sterile distilled, deionized water (ddH_2O) in a class II biological safety cabinet and unless otherwise specified. All reagents should minimally be analytical grade and weighed at room temperature. Here, we provide a protocol that is intended for research purposes only. Comply with all waste disposal regulations when disposing of waste material. All buffers should be autoclaved and at room temperature prior to use. All equipment should be cleaned and autoclaved following the directions of the manufacturer.

2.1 Reagents

1. Lyophilized human recombinant collagen (*see* **Note 1**).
2. 2-Morpholinoethane sulfonic acid monohydrate (MES) buffer: 0.05 M MES. De-gas buffer prior to use.
3. *N*-Hydroxysuccinimide (NHS) solution: 10 % (w/v) NHS in 0.05 M MES (*see* **Note 2**).
4. *N*-(3-Dimethylaminopropyl)-*N'*-ethylcarbodiimide (EDC) solution: 10 % (w/v) EDC in 0.05 M MES (*see* **Note 2**).
5. 0.1 M sodium phosphate buffer (Na_2HPO_4).
6. 0.1 M phosphate buffered saline (PBS).
7. Chloroform.

2.2 Equipment

1. T-Piece (Idex Health & Science, US, P-713-01).
2. Three pieces of female luers for T-piece (Idex Health & Science, US, P-624).
3. Two 2 mL glass syringes.
4. Rubber Septum (Restek Corp, 27169—17 mm).
5. Size 2 cork borer for cutting rubber septum (Fisher Scientific, 136–043306).
6. 50, 100, and 200 μL Hamilton microsyringes.
7. 10 mL Plastic sterile syringe with sterile caps.
8. 500 μm mold (*see* Fig. 2).
9. Mold jigs (*see* Figs. 3 and 4).
10. Sterile spatula.

3 Methods

All mixing should be performed at 4–6 °C. We recommend the use of a Pelletier cooling plate, but an ice water bath can also be used.

1. Prepare the collagen solution by dissolving the required amount of lyophilized collagen into sterile ddH_2O to achieve a solution of the desired percentage. The solid collagen content may range from 10 to 15 % w/w as required (*see* **Note 2**).
2. Transfer the collagen solution to a sterile 10 mL plastic syringe, seal this with sterile caps and centrifuge at 3,500 rpm for 30 min to remove entrapped air from the solution. Allow the syringe to rest, preferably overnight. Repeat the centrifugation step until no air bubbles are visible within the collagen solution.

Fig. 2 500 μm polypropylene molds

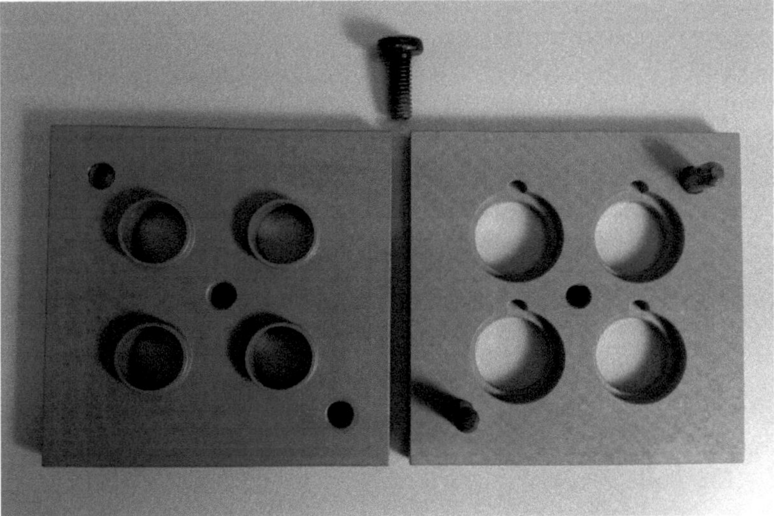

Fig. 3 Stainless steel jig assembly

3. Attach tubing, approximately 2 cm in length to the syringe tip. This tubing will be used to aid in the transfer of the collagen solution to the mixing syringe.
4. Introduce the plunger into the barrel of the syringe using a thin, sterile wire-like needle positioned alongside the plunger, and insert the plunger very slowly until the tip of this apparatus contacts the collagen solution in the syringe. Ensure that the needle is always positioned in the air pocket between the

Fig. 4 Jig assembly containing 500 μm molds prior to filling

plunger tip and the collagen to smoothly vent the air. It is critically important to avoid the introduction of any air into the syringe.

5. Slowly extract the needle.
6. Prepare the jigs and molds as shown in Figs. 3 and 4.
7. Assemble the T-pieces as shown in Fig. 5. Note that one port of the T-piece is sealed with a rubber septum to allow injection of cross-linking agents into the collagen solution without the introduction of air.
8. Load one glass syringe with MES buffer and attach to the T-piece. By pushing the MES through the T-piece ensure that the T-piece and glass syringe are air bubble free. Expel all the MES through the T-piece, leaving the plunger fully compressed and MES only remaining in the T-piece.
9. Load a second glass syringe with the required amount of collagen solution. Record the weight of the collagen solution and attach the syringe to the T-piece.
10. Return the collagen solution in the 10 mL plastic syringe to a 4–6 °C fridge.
11. Thoroughly mix the collagen solution with the MES buffer.
12. Check frequently to ensure that the two glass syringes are firmly attached to the T-piece.
13. Calculate the required amount of EDC and NHS. The amount of solution to be added is based on the molar equivalent ratio of EDC–NHS–ε-amine groups of collagen. In this protocol the ratio to be used is 0.4:0.4:1.0 (*see* **Notes 3** and **4**).

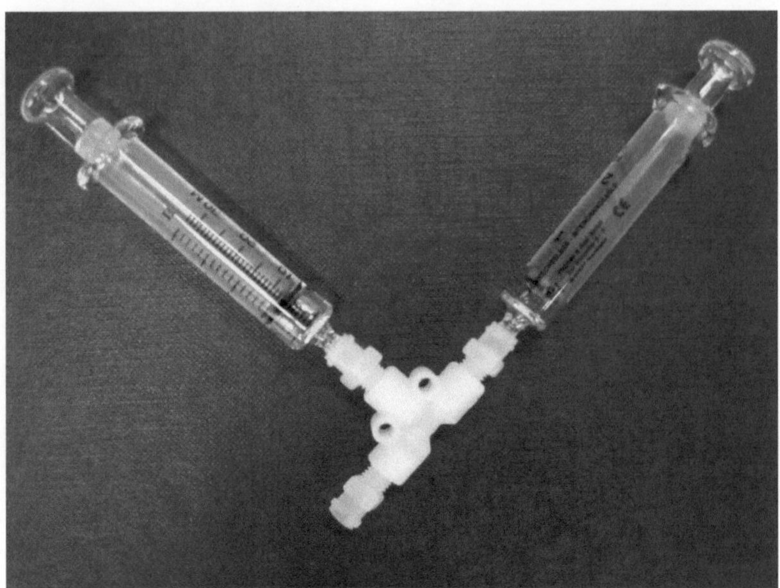

Fig. 5 T-piece mixing system with syringes attached

14. Inject the required amount of NHS solution through the septum and thoroughly mix with the collagen solution. Ensure that no air is introduced into the T-piece mixing system.
15. Inject the EDC solution through the septum and thoroughly mix with the collagen solution. Ensure that no air is introduced into the T-piece mixing system.
16. Push all the collagen mixture into one syringe and quickly detach this syringe from the T-piece mixing system.
17. Quickly dispense the required amount of collagen into the concave lower molds located in the lower jigs. Quickly place the top molds over the bottom molds in the jig to evenly distribute and seal the collagen solution within the mold (*see* **Notes 5** and **6**).
18. Attach the top of the jig and tighten bolts, but do not over tighten.
19. Place the jigs with the filled molds in a humidified chamber and place the chamber at room temperature for 24 h followed by 24 h at 37 °C.
20. Remove the molds from the jig and immerse the molds into sterile 0.1 M PBS for 24 h.
21. Carefully remove the corneal substitutes from molds using a spatula and transfer to individual vials.
22. Wash corneal substitutes with 0.1 M sodium phosphate for 3–4 h while gently shaking at 4–6 °C and then check implant integrity (*see* Fig. 6).
23. Remove the 0.1 M sodium phosphate and replace with sterile 0.1 M PBS. Exchange the PBS wash daily for 7 days.

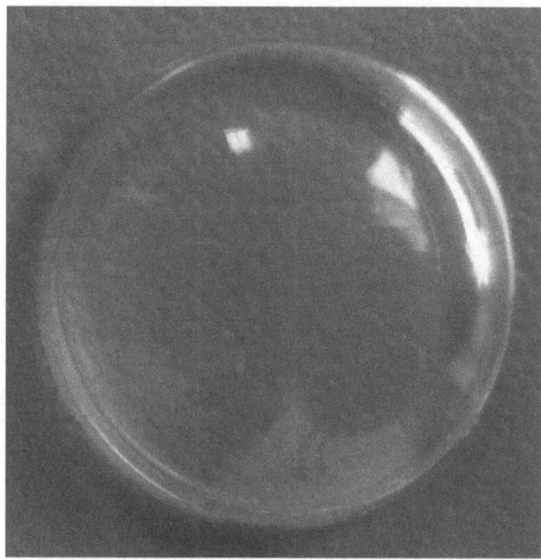

Fig. 6 Collagen-based corneal substitute

24. Place corneal substitutes in a final storage solution of 0.1 M PBS containing 1 % (v/v) chloroform.
25. Corneal substitutes should be stored at 4–6 °C until further use.

4 Notes

1. Prior to lyophilization, the pH of human recombinant collagen may require adjustment by dia-filtration in order that cross-linking occurs within the optimum pH range.
2. The collagen solution should be mixed for 3–7 days by mechanical or hand mixing until a homogeneous mixture is obtained. During this time, collagen should be stored at 4–6 °C.
3. To prepare the EDC and NHS solutions: Solutions should be prepared just prior to use. EDC is subject to hydrolysis and moisture must be avoided until immediately before use. All reagents must be at room temperature prior to weighing:

   NHS—20 mg in 200 μL of 0.05 M MES (10 % (w/v)).

   EDC—20 mg in 200 μL of 0.05 M MES (10 % (w/v)).
4. Always use a digital microbalance to weigh out the powder. Transfer the powder to a sterile, preferably electrostatic free weighing vials and vortex solution for 30 s. All solutions should be stored at 4–6 °C.
5. Cross-linking occurs rapidly and therefore the collagen solution must be dispensed quickly into the molds.
6. We use contact lens molds. Custom jigs that fit the molds are best. If those are not available, molds can be clamped between two sheets of metal plating or glass. For research purposes,

hydrogels can also be cast between two glass plates separated by teflon spacers. The glass or metal plating has to be clamped together tightly to make a hydrogel that is crosslinked and will not swell a lot.

Acknowledgments

We thank the Natural Sciences and Engineering Research Council of Canada, Canadian Stemcell network for research funding (M.G.). We thank our colleagues, Drs. David Carlsson and Yuwen Liu, for their contributions to the development of early constructs.

References

1. Merrett K, Fagerholm P, McLaughlin CR, Dravida S, Lagali N, Shinozaki N et al (2008) Tissue-engineered recombinant human collagen-based corneal substitutes for implantation: performance of type I versus type III collagen. Invest Ophthalmol Vis Sci 49:3887–3894
2. Yang C, Hillas PJ, Baez JA, Nokelainen MJ, Balan J, Tang J et al (2004) The application of recombinant human collagen in tissue engineering. BioDrugs 18:103–119
3. Stein H, Wilensky MI, Tsafrir Y, Rosenthal M, Amir R, Avraham T et al (2009) Production of bioactive, post-translationally modified, heterotrimeric, human recombinant type-I collagen in transgenic tobacco. Biomacromolecules 10:2640–2645
4. Kew SJ, Gwynne JH, Enea D, Abu-Rub M, Pandit A, Zeugolis D et al (2011) Regeneration and repair of tendon and ligament tissue using collagen-fibre biomaterials. Acta Biomater 7:3237–3247
5. Williams DF (1985) Biocompatibility of tissue analogs, vol 1. CRC, Boca Raton, FL
6. Sionkowska A (2011) Current research on the blends of natural and synthetic polymers as new biomaterials: review. Prog Polym Sci 36:1254–1276
7. Nam K, Kimura T, Funamoto S, Kishida A (2010) Preparation of a collagen/polymer hybrid gel designed for tissue membranes. Part I: controlling the polymer-collagen cross-linking process using an ethanol/water co-solvent. Acta Biomater 6:403–408
8. Stricklin GP, Hibbs MS, Nimmi ME (eds) (1988) Collagen. CRC, Boca Raton, FL
9. Gendler E, Gendler S, Nimni ME (1984) Toxic reactions evoked by glutaraldehyde-fixed pericardium and cardiac valve tissue bioprostheses. J Biomed Mater Res 18:727–736
10. Duncan AC, Boughner D, Vesley I (1996) Dynamic glutaraldehyde fixation of a porcine aortic valve xenograft: I. Effect of fixation conditions on the final tissue viscoelastic properties. Biomaterials 17:1849–1856
11. Nakajima N, Ikada Y (1995) Mechanism of amide formation by carbodiimide for bioconjugation in aqueous media. Bioconjug Chem 6:123–130
12. Gilles MA, Hudson AQ, Borders C (1990) Stability of water soluble carbodiimides in aqueous solution. Anal Biochem 184:244–248

Chapter 11

Fabrication of a Corneal-Limbal Tissue Substitute Using Silk Fibroin

Laura J. Bray, Karina A. George, Shuko Suzuki, Traian V. Chirila, and Damien G. Harkin

Abstract

Fibroin extracted from silkworm cocoon silk provides an intriguing and potentially important biomaterial for corneal reconstruction. In this chapter we outline our methods for producing a composite of two fibroin-based materials that support the cocultivation of human limbal epithelial (HLE) cells and human limbal stromal (HLS) cells. The resulting tissue substitute consists of a stratified epithelium overlying a three-dimensional arrangement of extracellular matrix components (principally "degummed" fibroin fibers) and mesenchymal stromal cells. This tissue substitute is currently being evaluated as a tool for reconstructing the corneal limbus and corneal epithelium.

Key words Silk fibroin, Corneal-limbus, Epithelium, Mesenchymal stromal cells

1 Introduction

Fibroin is the main structural protein found in silk fibers [1]. The best-characterized fibroin is that produced by larvae of the domesticated silk moth *Bombyx mori* [1, 2]. In its native state, *B. mori* fibroin is polymerized into filaments and bundled together with a glue-like protein known as sericin. While silk has been widely used in the past as a surgical suture, it is now largely replaced by synthetic materials owing to cases of immune responses to native silk [3–5]. This adverse reaction to silk is attributed primarily to sericin [1]. Removal of sericin from silk, however, can be achieved via a simple "degumming" process, thus releasing the fibroin fibers and in doing so providing a potentially valuable biomaterial for a range of biomedical applications [6, 7].

Several characteristics of fibroin suggest that this material could provide a useful alternative to amniotic membrane (AM) as a biomaterial for corneal reconstruction. Fibroin can be readily purified in large quantities and at low cost from *B. mori* cocoons. The resulting product, in the form of either degummed fibers or

Bernice Wright and Che J. Connon (eds.), *Corneal Regenerative Medicine: Methods and Protocols*, Methods in Molecular Biology, vol. 1014, DOI 10.1007/978-1-62703-432-6_11,

an aqueous solution, can be readily fashioned into a variety of structures including transparent membranes, fibrous mats and sponges [8]. The transparency of fibroin membranes (>96 % across visible spectrum) [9] is similar to that achieved using collagen, but superior to that of AM. Fibroin mats and sponges display poor transparency (20–25 %) and thus are unsuitable for repairing the corneal stroma, but retain potential as scaffolds for limbal and scleral cells [10].

In general, fibroin-based materials display good mechanical properties, a slow rate of biodegradation and are well tolerated when implanted into living tissues including the cornea [11, 12]. Perhaps the most useful feature of fibroin, however, may be the mechanism by which this material interacts with the cell surface. In the absence of recognizable cell-binding motifs, *B. mori* fibroin is essentially inert, but acceptable levels of cell attachment, growth and differentiation can be obtained via addition of exogenous factors [9, 10, 12–14]. Fibroin based materials can potentially be "tuned" according to the cell type and desired outcome. By comparison, the materials present in AM including collagen, present inherent "instructions" to cells via engagement with integrin receptors that, whilst often beneficial, are unlikely to be optimal for every cell type or application.

We presently outline the methods used in our laboratory to construct a synthetic tissue substitute consisting of two fibroin-based materials and primary cultures of cells isolated from the human corneal limbus (Fig. 1). The first fibroin component is a 3D mat (~0.5 mm in thickness) of degummed fibroin fibers (Fig. 2f) that serves as a scaffold for human limbal stromal (HLS) cells, and the second component is a transparent fibroin membrane (~5 μm in thickness) that provides a substrate for human limbal epithelial (HLE) cells (Fig. 2c, e). Six critical steps are described in detail; (1) initial degumming of raw cocoon silk, (2) preparation of fibroin mats (stromal cell scaffold), (3) preparing an aqueous solution of fibroin from degummed fibers, (4) preparation of transparent fibroin membranes (epithelial cell substrate), (5) construction of the fibroin composite material, and (6) cultivation of limbal cells on and within the fibroin composite. The final product (Fig. 3) has yet to be used clinically, but preparations are currently underway to conduct a preclinical trial of safety and efficacy in our laboratory.

2 Materials

2.1 Processing of Cocoon Silk

1. Dried *B. mori* cocoons (Tajima Shoji Co. Ltd., Yokohama, Japan, *see* **Note 1**).
2. Surgical dressing scissors.
3. Sodium carbonate.

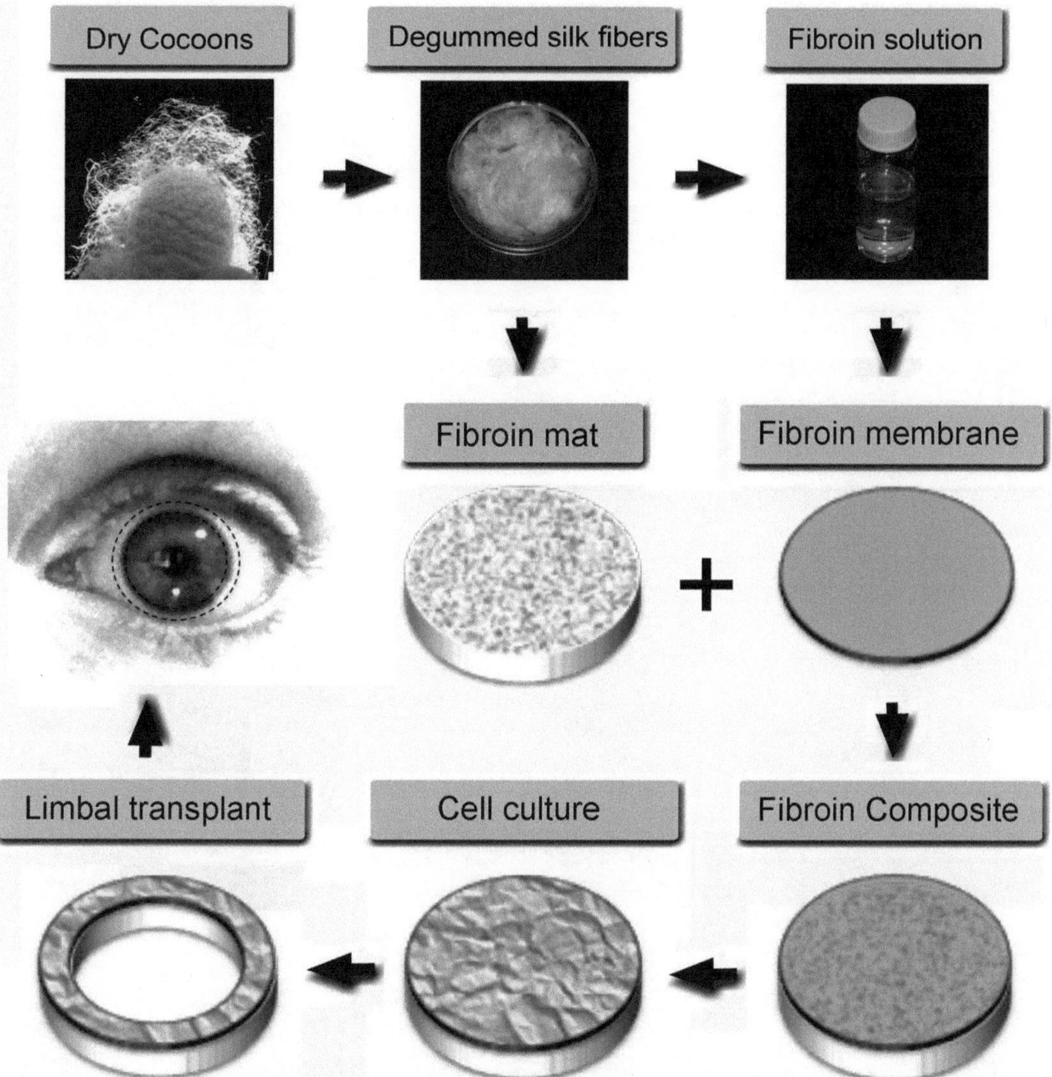

Fig. 1 Summary of steps used to produce a corneal-limbal tissue substitute from silk fibroin and cultured cells. The final product ("Limbal Transplant") can be cut to size using a trephine blade of required size at time of surgery. While the product illustrated is designed to repair the corneal limbus, the technique can be extended to include the central corneal epithelium by using a fibroin composite consisting of an intact fibroin membrane attached to a precut donut shape of fibroin mat

4. Ultrahigh purity (UHP) water (prepared to 18.2 MΩ cm at 25 °C).
5. Glass beakers (2 × 1 L and 1 × 25 mL capacities).
6. Watch glasses (large enough to cover beakers).
7. Hot plate/stirrer equipped with thermocouple.
8. Spatulas (2 × 150 mm Chattaway pattern).

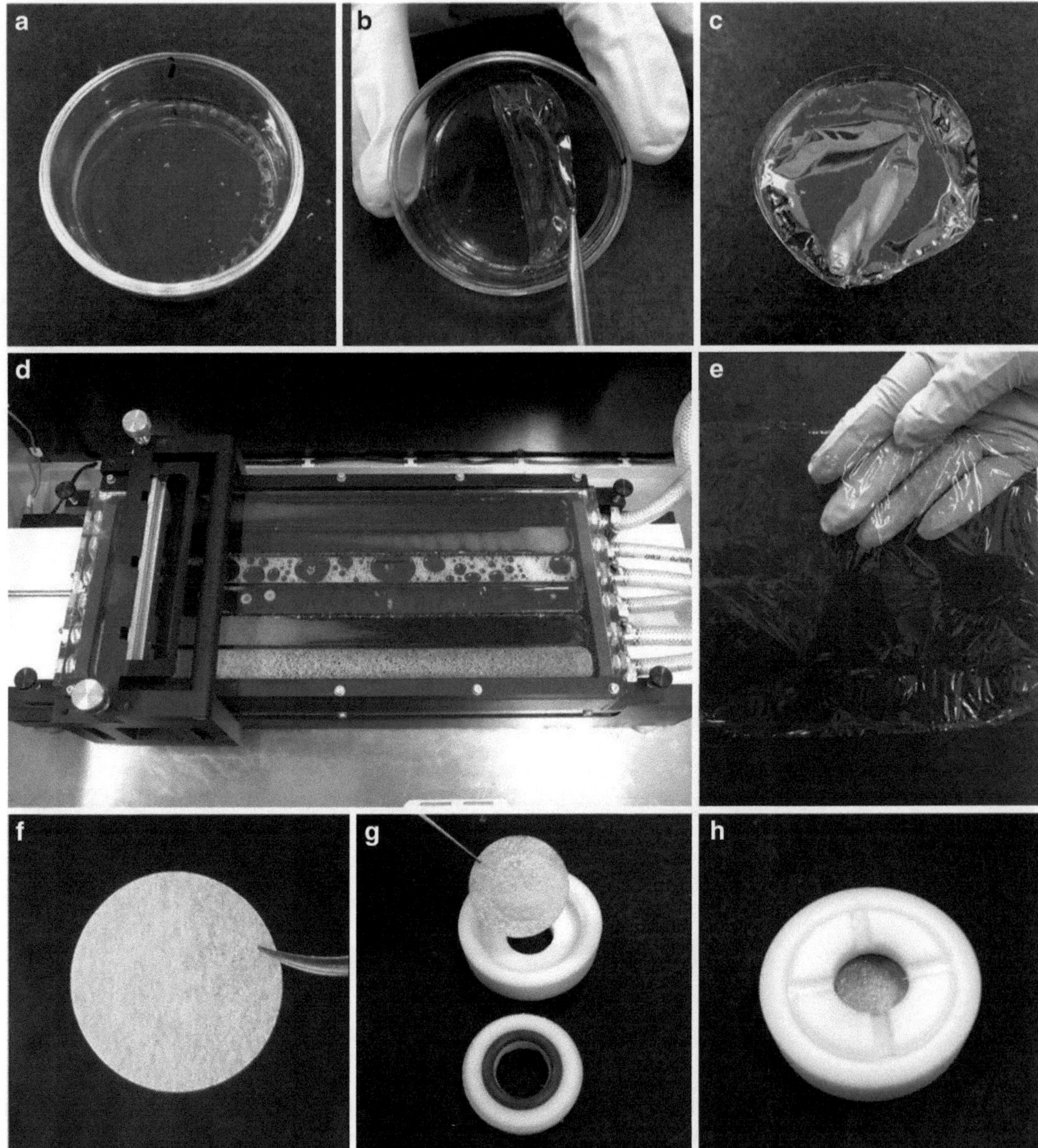

Fig. 2 Illustration of fibroin-based materials used. (**a**–**c**) Appearance of fibroin membranes immediately after casting and drying in a glass Petri dish, during removal and following removal from the Petri dish, respectively. (**d**) Casting table as viewed from above. The doctor blade mounted on a mechanical stage is located at the left hand side of the table. Air bubbles can be seen within the circulating water channels immediately beneath the glass plate. (**e**) Example of large fibroin membrane produced via the casting table method. (**f**) Appearance of dry fibroin mat. (**g**) Insertion of fibroin composite into Teflon® cell culture chamber. A silicon "o" ring can be seen within the inverted upper compartment prior to screwing down into the base. (**h**) Assembled fibroin composite ready for cell culture

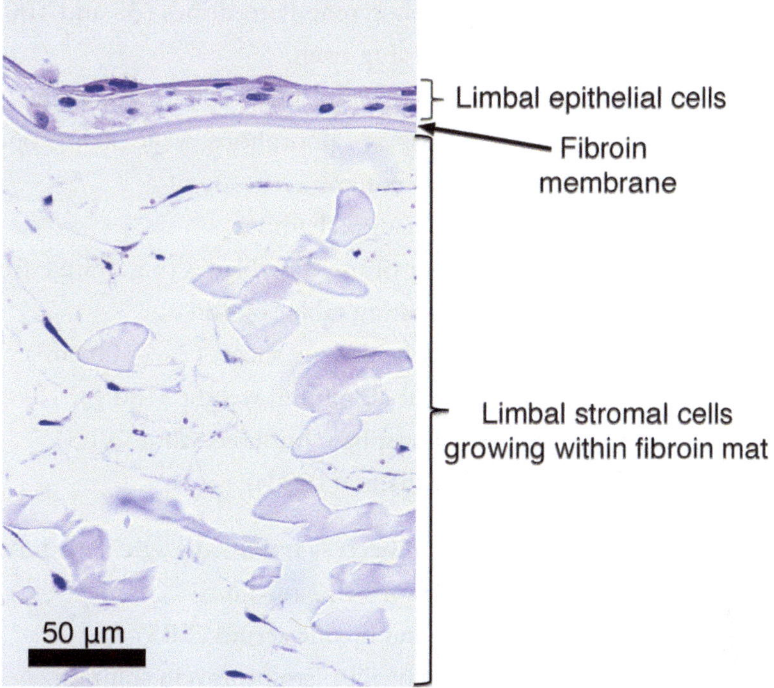

Fig. 3 Photomicrograph illustrating the cross-sectional structure of the final corneal-limbal tissue substitute constructed from silk fibroin, human limbal epithelial cells, and human limbal stromal cells (4 μm paraffin section stained with Ehrlich's hematoxylin and eosin). With further refinements to the fibroin membrane (e.g., folding and surface functionalization) it may be possible to more closely mimic the architecture and chemical composition of the limbal stem cell niche (i.e., limbal crypts)

9. Fume hood.
10. Forceps (2 × 16 cm surgical dressing forceps and 2 × No. 7 watchmaker forceps).
11. Plastic centrifuge tube with cap (50 mL capacity).
12. Formic acid (98–100 %).
13. Flask shaker.
14. Lithium bromide (LiBr).
15. Glass water bath (diameter 115 mm, height 65 mm).
16. Plastic 50 mL syringes.
17. Syringe needles (18 G).
18. Syringe filters (0.8 and 0.2 μm pore sizes).
19. Dialysis cassettes (30 mL capacity).
20. Magnetic stirrer bars.
21. Small plastic storage container (30–50 mL).

22. Polystyrene Petri dishes (35 and 100 mm diameter).
23. Drying oven.
24. Glass Petri dishes (60 mm diameter)
25. Topas® (amorphous cyclic olefin copolymer).
26. Cyclohexane.
27. Vacuum chamber.
28. Set of trephine blades (e.g., range from 6 to 35 mm).
29. Casting table (optional).
30. Polyethylene sheets (2× ~100 cm^2).
31. Flat aluminum weight (400 g).
32. Phosphate buffered saline (PBS).
33. Ethanol (100 %).

2.2 Cell Culture Reagents

1. Dulbecco's modified Eagle's medium.
2. Ham's F12 medium.
3. Fetal bovine serum (FBS).
4. Penicillin/streptomycin solution.
5. Epidermal growth factor.
6. Non-essential amino acids.
7. L-Glutamine.
8. 3,3,5-Triiodo-L-thyronine sodium salt (T3).
9. Adenine.
10. Transferrin.
11. Hydrocortisone.
12. Insulin
13. Isoproterenol (isoprenaline).
14. Murine 3T3 fibroblasts (ATCC, Cat. No. CCL-92).

3 Methods

3.1 Degumming of Raw Cocoon Silk

1. Cut dried cocoons into approximately 0.8 cm^2 pieces (approximately 16 pieces per cocoon) using scissors.
2. Weigh out 2.5 g of cocoon pieces.
3. Weigh out 2.12 g of sodium carbonate and add to 1 L of UHP water in a glass beaker.
4. Cover the beaker with a watch glass and heat the sodium carbonate solution to boiling point using a hot plate.
5. Add the weighed cocoon fragments to the boiling solution and use a spatula to disperse the silk fibers.

6. Boil cocoon fragments for 1 h.
7. Discard the solution and squeeze the degummed silk (fibroin) to remove as much liquid as possible (*see* **Note 2**). Rinse the fibroin twice with approximately 20 mL of UHP water.
8. Place the fibroin in a beaker filled with 1 L of 60–70 °C UHP water for 20 min. Cover the beaker with a watch glass and stir the fibroin occasionally using a spatula.
9. Repeat **steps 7–8** until the fibroin has been rinsed three times and has been squeezed dry.
10. Loosen the resulting fibrous mat by pulling the fibers apart gently (*see* **Note 3**).
11. Put the expanded mat into a fume hood to dry for at least 12 h.
12. The degummed fibers can now be used to prepare fibrous mats or can be processed further to produce a silk fibroin solution used in the fabrication of a fibroin membrane.

3.2 Preparation of Fibrous Mats (Stromal Cell Scaffold)

1. Separate the degummed fibers obtained above by gently teasing them apart using dressing forceps.
2. Place the dispersed fibers (0.25 g) in a plastic centrifuge tube with 25 mL of formic acid and shake for 20 min using a flask shaker on lowest speed.
3. Pour the resulting solution into a polystyrene Petri dish (100 mm diameter) and slowly dry over several days at room temperature.
4. Detach the dried fibrous mat from the Petri dish using watchmaker forceps and cut into circles of required size (Fig. 2f) using a trephine blade.
5. Store the circles of fibroin mat in a clean Petri dish.

3.3 Preparation of Fibroin Solution

1. Weigh the degummed silk fibers.
2. Determine the volume of 9.3 M LiBr solution required to create a 10 % (w/v) solution of fibroin by using the formula: volume (mL) = 10 × mass of fibroin (*g*).
3. Weigh out the required mass of LiBr into a 25 mL glass beaker: mass (g) = 9.3 × volume (mL; calculated in **step 2**) × (86.85/1,000).
4. Dissolve the LiBr in UHP water to approximately 2 mL less than the volume calculated in **step 2**. Mix the solution to ensure homogeneity (*see* **Note 4**).
5. Put the beaker containing the LiBr solution into the glass water bath (half filled with tap water) and heat to 60 °C using the hotplate and thermocouple (probe should be put into the water within the water bath, not directly into the LiBr solution). Put a watch glass over the beaker to slow evaporation.

6. When the LiBr solution has reached the required temperature, add the fibroin and incubate for 4 h (*see* **Note 5**).
7. Monitor the volume of the fibroin solution over the 4-h period and, if necessary, adjust the volume back to that calculated in **step 2** by drop-wise addition of UHP water.
8. Remove the solution from the water bath and allow this to cool down for 10 min on the bench.
9. Slowly load the solution into a 50 mL plastic syringe using an 18 G needle. Slowly filter the solution through the 0.80 and 0.20 μm syringe filter (connected in series) directly into the dialysis cassette (presoaked in UHP water) via an 18 G needle (fitted on to the end of a 0.20 μm syringe filter).
10. Put the filled dialysis cassette into a beaker containing 1 L of UHP water, gently stirred with a magnetic stirrer bar.
11. Change the water at 1, 4, 7, 18, 30, and 48-h intervals after the start of the dialysis procedure (*see* **Note 6**).
12. After 72 h, slowly remove the solution from the dialysis membrane using an 18 G needle fitted onto a 50 mL plastic syringe.
13. Slowly filter the solution through the 0.80 and 0.20 μm syringe filters (connected in series) directly into a plastic storage vessel (*see* **Note 7**).
14. To calculate the concentration of fibroin, first weigh a 35 mm diameter plastic Petri dish (with lid removed) and then add 1 mL of dialyzed fibroin solution to the dish.
15. Cover the dish and place into the drying oven at 60 °C overnight.
16. Let the dish cool down and then reweigh.
17. Calculate the concentration of fibroin (% (w/v)) based upon the difference in dry weight of the dish before and after addition of the fibroin solution (typically between 3.4 and 3.7 %).
18. Store the solution in a sealed plastic container at 4 °C and use within 1 month or until the solution appears cloudy (*see* **Note 8**).

3.4 Preparation of Fibroin Membrane (Epithelial Cell Substrate)

Option 1: Petri Dish Method

1. Calculate the volume of fibroin solution required per 60 mm glass Petri dish based upon the desired thickness (*see* **Note 9**).
2. Pre-coat the Petri dish with a film of amorphous cyclic olefin copolymer (Topas®) by dispensing 0.7 g of the copolymer dissolved in 10 mL of cyclohexane evenly across the dish surface and allowing the solvent to evaporate slowly overnight (*see* **Note 10**).

3. In order to produce a fibroin membrane of approximately 6 μm in thickness, dispense 1.68 mL of a 1.78 % (w/v) fibroin solution onto the Topas®-coated dish and dry overnight.
4. Using watchmaker forceps, gently remove the fibroin membrane from the Petri dish as a single sheet while still attached to the copolymer film (Fig. 2b).
5. "Water anneal" (i.e., stabilize the fibroin) by treatment at −80 kPa in a vacuum chamber for 6 h at room temperature in the presence of approximately 100 mL of water in a container.
6. Peel the fibroin membrane away from the copolymer film.
7. Cut the membrane into circles of desired size using a trephine blade.

Option 2: Casting Table Method (*see* **Note 11**)

1. Adjust the doctor blade to a height of 0.4 mm from the glass surface.
2. Cast a film of amorphous cyclic olefin copolymer (Topas®) onto the glass surface by applying 1.4 g of copolymer in 20 mL of cyclohexane and allow the solvent to evaporate slowly (*see* **Note 10**).
3. Apply 15 mL of a 1.78 % (w/v) fibroin solution onto approximately 200 cm^2 of the copolymer-coated glass surface and allow to dry overnight.
4. Clean the doctor blade directly after casting each solution with the appropriate solvent.
5. Remove the fibroin membrane from the glass as a single sheet while attached to the copolymer film and water anneal (i.e., stabilize the fibroin) as described above for the Petri dish method (first **step 5** in Subheading 3.4).
6. Peel the fibroin membrane away from the copolymer membrane.
7. Store the sheets of fibroin membrane (Fig. 2e) between sheets of paper within plastic sheet protectors.
8. As required, remove fibroin membranes from storage and cut the membrane into circles of desired size using a trephine blade.

3.5 Creation of Fibroin Composite

By combining the mat of degummed silk fibers with an overlying sheet of fibroin membrane, a composite material is produced to enable cocultivation of epithelial cells and stromal cells (*see* **Note 12**).

1. Coat one surface of the dried fibroin mat with a 1 % (w/v) aqueous fibroin solution and then carefully layer a sheet of fibroin membrane onto this surface.
2. Place the composite fibroin material between two sheets of polyethylene and apply an aluminum weight on top.
3. After 24 h, remove the weight and allow the composite scaffold to dry over several days.

4. Once dry, water anneal the freshly dried fibroin within the composite scaffold as described above in the first **step 5** of Subheading 3.4.
5. Hydrate the composite scaffolds in PBS.
6. Sterilize the scaffolds by submersion in 70 % (v/v) ethanol for 45 min, wash three times with sterile PBS, and store in PBS until use.
7. The sterilized composite can then be placed directly into a culture dish (e.g., 6-well culture plate) or mounted within a supporting culture chamber (*see* **Note 13** and Fig. 2g, h).

3.6 Cell Culture

Using primary cultures of human limbal epithelial (HLE) cells (*see* **Note 14**) and human limbal stromal (HLS) cells (*see* **Note 15**) a limbal tissue substitute can be generated by seeding these cells either onto or into the fibroin membrane and fibroin mat respectively (*see* **Note 16**). All procedures are performed in a class II biological safety cabinet and cultures are maintained at 37 °C in a humidified incubator supplied with 5 % CO_2 in air.

1. To begin, invert the sterile composite fibroin scaffold so that the fibrous mat is facing upwards and seed freshly isolated HLS cells at a density of 10^5 cells/cm^2 in DMEM medium supplemented with 10 % (w/v) FBS, 2 mM l-glutamine and antibiotic solution.
2. Culture the HLS cells within the fibrous mat for approximately 20–24 h to allow settling and attachment within the 3D scaffold.
3. Revert the composite fibroin scaffold so that the membrane component is now orientated upwards and seed freshly isolated HLE cells at a density of 10^5 cells/cm^2 in epithelial growth medium (*see* **Note 14**).
4. Maintain the HLE/HLS coculture for approximately 14 days in epithelial growth medium.
5. For routine histology, the resulting limbal tissue substitute can be fixed for 1–2 h in neutral buffered formalin (3.6 % (w/v) formaldehyde) and processed into paraffin for sectioning and staining with hematoxylin and eosin (*see* **Note 17** and refer to Fig. 3).

4 Notes

1. Techniques for processing cocoon silk fibroin can be significantly affected by choice of species and supplier. The domesticated silkworm *B. mori* is the best-characterized species and reproducible results are obtained using cocoons from Tajima Shoji Co. Ltd.

2. Use two spatulas to squeeze out liquid from the fibroin during the first wash since the fibroin will be extremely hot coming directly from the heated vessel. Gloved hands can then be used during subsequent washes.
3. We find that it is best not to pull the fibers apart leaving gaping holes. Try to gently pull apart the fibers in bulky areas in order for the mat to dry fully.
4. This step needs to be performed with caution since dissolution of LiBr in UHP water is very exothermic. We find that it is best to make up fresh LiBr solution for use on a given day. We do not stir the water bath or the LiBr solution.
5. Once the fibroin is added to the LiBr solution, it should start to dissolve fairly quickly and should be completely dissolved within approximately 2-h. The solution may initially appear yellow to orange in color and change to a more peach-like color by the end of the 4-h period.
6. It is not critical to change the water at the exact times given here, but as close as possible to those times is recommended.
7. If the solution appears cloudy, then it should be refiltered. A needle is not required on the end of the 0.20 μm filter to transfer the solution into a plastic storage vessel.
8. Care should be taken when handling and storing fibroin solutions as agitation caused by shaking or shear forces promote gel formation.
9. We routinely make fibroin membranes that are between 5–8 μm in thickness. In 60 mm glass Petri dishes, we calculate 0.28 mL of solution per μm membrane thickness. For example, 1.68 mL of fibroin solution (1.78 % (w/v)) is used to make a 6 μm thick membrane.
10. It is important that the Topas® dry slowly otherwise an uneven texture will be produced that subsequently affects the architecture of the fibroin membrane. To achieve a slow rate of drying we either partially cover the Petri dish with its lid or, in the case of the casting table method, we cover the plate by resting an upturned plastic container on the mounting brackets and close the lid to the box in which the apparatus is housed.
11. While suitable test membranes can be produced in Petri dishes, a casting table enables production of larger quantities of fibroin membrane, with greater control over membrane thickness. We utilize a custom-made film-casting table consisting of an optically flat sheet of glass mounted below a doctor blade that moves across the glass sheet at a constant height from the glass surface (Fig. 2d). An even temperature across the table surface is facilitated by pumping water through three "U" shaped channels within the aluminum block on which the glass rests.

Blade height is adjusted manually with the aid of two digital indicators placed at either end. Blade speed (~2.5 cm/s) is controlled using a programmable stepper motor.

12. An alternative design that might be used to repair the central corneal epithelium as well as the surrounding limbus is to use a composite produced by combining a complete circle of fibroin membrane with a precut fibroin mat (i.e., donut shape). The fibroin mat is unsuitable for repairing the central corneal stroma owing to poor transparency.

13. We prefer to mount our fibroin-based cell culture substrates within a custom-made holding chamber constructed from two screw-locking Teflon® rings and a silicon "o" ring (Fig. 2g, h). This technique can be used for membranes alone although care must be taken not to tear or puncture the membranes during assembly.

14. While a number of techniques and growth media are available for limbal epithelial cells, it is essential to use cultures that contain a high number of poorly differentiated cells. We therefore prefer the traditional technique of establishing and passaging cultures in the presence of growth-arrested murine 3T3 fibroblasts as first established for skin keratinocytes [15, 16]. The culture medium consists of a 1:3 mixture of Dulbecco's modified Eagle's medium (DMEM) and Ham's F12 medium, supplemented with 10 % (w/v) FBS, 1 % (v/v) penicillin/streptomycin solution, 10 ng/mL epidermal growth factor, 1 % (v/v) nonessential amino acids, 2 mM l-glutamine, 6.8 mg 3,3,5-triiodo-l-thyronine sodium salt (T3), 180 mM adenine, 5 mg/mL transferrin, 0.4 mg/mL hydrocortisone, 1 mg/mL insulin, and 10^{-5} M isoproterenol. The 3T3 fibroblasts are simply maintained in serum-supplemented growth medium and are growth-arrested immediately prior to seeding into epithelial cultures by treatment with gamma radiation (2 × 25 Gy).

15. Techniques for the characterization and cultivation of limbal stromal cells are an emerging field. Robust cultures can be readily established using serum-supplemented growth medium [17, 18]. Using this technique the majority of cells can be classified as mesenchymal stromal cells (MSC) by virtue of being immuno-positive for CD73, CD90, and CD105, and negative for CD34 and CD45 [17, 18].

16. As classic cell-adhesion motifs such as the RGD peptide are absent in *B. mori* silk fibroin, the mechanism of cell attachment to structures fabricated from this protein is largely dependent upon the addition of exogenous cell attachment factors. In the

case of HLE cells and HLS cells, the cell attachment factors present in serum-supplemented growth medium have been found to be sufficient although there remains significant opportunity for optimization.

17. Great care must be taken during tissue processing and embedding to ensure that the epithelial surface is not unduly damaged. A number of techniques may assist with protection including immersion of the fixed culture in an agar gel or wrapping in cigarette paper prior to placing into a tissue processing cassette. We also recommend truncating the processing times significantly compared with those used for standard tissue. It is useful to apply a drop of hematoxylin solution to the fixed culture prior to processing so that paraffin sections can be rapidly screened for areas of interest without the need for de-paraffinization and complete staining. During embedding, we cut the culture through the middle to create two semicircles and orientate the two cut ends facing downwards (i.e., closest to block surface during microtomy). Highly adhesive glass slides (e.g., SuperFrost Ultra Plus®, Menzel-Glaser) should be used to facilitate binding of sections and it is advantageous to mount multiple sections per slide in case some sections are still lost (especially for immunostaining since this process involves multiple washing steps). Finally, we recommend trialing different section thicknesses (e.g., 3–6 μm) to improve section attachment and final appearance.

Acknowledgments

This work was supported in part by a grant from the National Health & Medical Research Council of Australia (NH&MRC Project Grant 553038) with supplementary funding from the former Discipline of Medical Sciences (Queensland University of Technology, Brisbane, Australia) and the Prevent Blindness Foundation, Queensland, Australia (supported through Viertel's Vision). The casting table illustrated and described in this chapter (Fig. 2d) was designed and manufactured by Armin Liebhardt, William Gordon and colleagues from the Design and Manufacturing Centre, Science and Engineering Faculty, Queensland University of Technology, Brisbane, Australia. Finally, we wish to acknowledge the advice and encouragement received from Peter Madden and Lawrie Hirst (Queensland Eye Institute), Ivan Schwab (UC Davis), Dietmar Hutmacher (Queensland University of Technology), David Kaplan (Tufts University), and Brian Lawrence (Cornell University).

References

1. Altman GH, Diaz F, Jakuba C, Calabro T, Horan RL, Chen J et al (2003) Silk-based biomaterials. Biomaterials 24:401–416
2. Marsh RE, Corey RB, Pauling L (1955) An investigation of the structure of silk fibroin. Biochim Biophys Acta 16:1–34
3. Moore TE Jr, Aronson SB (1969) Suture reaction in the human cornea. Arch Ophthalmol 82:575–579
4. Salthouse TN, Matlaga BF, Wykoff MH (1977) Comparative tissue response to six suture materials in rabbit cornea, sclera, and ocular muscle. Am J Ophthalmol 84:224–233
5. Soong HK, Kenyon KR (1984) Adverse reactions to virgin silk sutures in cataract surgery. Ophthalmology 91:479–483
6. Murphy AR, Kaplan DL (2009) Biomedical applications of chemically-modified silk fibroin. J Mater Chem 19:6443–6450
7. Wang Y, Kim HJ, Vunjak-Novakovic G, Kaplan DL (2006) Stem cell-based tissue engineering with silk biomaterials. Biomaterials 27:6064–6082
8. Harkin DG, George KA, Madden PW, Schwab IR, Hutmacher DW, Chirila TV (2011) Silk fibroin in ocular tissue reconstruction. Biomaterials 32:2445–2458
9. Bray LJ, George KA, Ainscough SL, Hutmacher DW, Chirila TV, Harkin DG (2011) Human corneal epithelial equivalents constructed on Bombyx mori silk fibroin membranes. Biomaterials 32:5086–5091
10. Bray LJ, George KA, Hutmacher DW, Chirila TV, Harkin DG (2012) A dual-layer silk fibroin scaffold for reconstructing the human corneal limbus. Biomaterials 33:3529–3538
11. Meinel L, Hofmann S, Karageorgiou V, Kirker-Head C, McCool J, Gronowicz G et al (2005) The inflammatory responses to silk films in vitro and in vivo. Biomaterials 26:147–155
12. Higa K, Takeshima N, Moro F, Kawakita T, Kawashima M, Demura M et al (2011) Porous silk fibroin film as a transparent carrier for cultivated corneal epithelial sheets. J Biomater Sci 22:2261–2276
13. Madden PW, Lai JN, George KA, Giovenco T, Harkin DG, Chirila TV (2011) Human corneal endothelial cell growth on a silk fibroin membrane. Biomaterials 32:4076–4084
14. Shadforth AM, George KA, Kwan AS, Chirila TV, Harkin DG (2012) The cultivation of human retinal pigment epithelial cells on Bombyx mori silk fibroin. Biomaterials 33:4110–4117
15. Rheinwald JG, Green H (1975) Serial cultivation of strains of human epidermal keratinocytes: the formation of keratinizing colonies from single cells. Cell 6:331–343
16. Rheinwald JG, Green H (1977) Epidermal growth factor and the multiplication of cultured human epidermal keratinocytes. Nature 265:421–424
17. Ainscough SL, Linn ML, Barnard Z, Schwab IR, Harkin DG (2011) Effects of fibroblast origin and phenotype on proliferative potential of limbal epithelial progenitor cells. Exp Eye Res 92:10–19
18. Bray LJ, Heazlewood CF, Atkinson K, Hutmacher DW, Harkin DG (2012) Evaluation of methods for cultivating limbal mesenchymal stromal cells. Cytotherapy 14:936–947

Chapter 12

Cultivation of Limbal Epithelial Cells on Electrospun Poly (lactide-*co*-glycolide) Scaffolds for Delivery to the Cornea

Pallavi Deshpande, Charanya Ramachandran, Virender S. Sangwan, and Sheila MacNeil

Abstract

In delivering tissues to the body, both natural and synthetic materials have been used. Currently, a natural membrane, the human amniotic membrane (AM), is used to deliver limbal epithelial cells (LEC) to the cornea. AM presents inherent problems with structural variation and requires extensive serological screening before use. Therefore alternatives are required to improve the predictability in clinical outcomes and economic costs associated with the use of this biological substrate. In this chapter, we describe the development of an alternative, structurally simple, synthetic biodegradable electrospun scaffold based on poly(lactide-*co*-glycolide) (PLGA: materials used in dissolvable sutures) to replace AM.

Key words Electrospinning, Poly(lactide-*co*-glycolide), Limbal epithelial cell cultivation

1 Introduction

The most common method for delivering LEC to the cornea is currently by culturing these cells on donor AM [1–14]. Human AM, which lacks immunogenicity [4, 5], provides LEC with a basement membrane which is similar in structure to that of the cornea [6]. AM also preserves the stemness of cells [7, 8], and presents anti-inflammatory and antimicrobial properties [9–11]. Although the procedure for using this biological substrate to deliver cells has been successful clinically [12], this tissue is derived from a human source. The use of human AM is therefore donor dependent, and inter- and intra-donor variations in the outcome of using this tissue for treatment purposes are often presented [13]. In addition to this, all human donor tissue needs to be screened for human immunodeficiency virus and other viral diseases (hepatitis B and C) before their use [14]. This requires that tissues are banked according to best-practice tissue banking, and they must also be stored and processed accordingly. For all of these reasons, a great

Bernice Wright and Che J. Connon (eds.), *Corneal Regenerative Medicine: Methods and Protocols*, Methods in Molecular Biology, vol. 1014, DOI 10.1007/978-1-62703-432-6_12,

deal of work is now being carried out to develop a synthetic alternative to AM [15–19].

The field of tissue engineering provides many opportunities to develop substrates that are appropriate for supporting cell growth and that are biodegradable, i.e., replaced by tissues in the body. Substrates must always be designed with respect to their specific application. AM is a sacrificial substrate because it does not survive long-term (sutured or glued to the cornea), and essentially breaks down over several weeks or months, and allows LEC to be delivered to the cornea. Thus, for our application, we require a substrate that will support LEC attachment and proliferation, but that will break down relatively rapidly on the cornea and let LEC be transferred to the corneal surface. Here, we describe the fabrication of electrospun scaffolds using biodegradable PLGA polymers and subsequent culture of LEC on these biomaterials. Electrospinning is a simple and controllable technique resulting in a scaffold of desired fiber diameter and thickness depending on the polymer used, by varying the fabrication parameters (voltage, flow rate, surrounding temperature) as well as the polymer concentration (*see* Chapter 13) [16].

By considering these factors, the degradation of PLGA scaffolds as a whole may be controlled. These synthetic off-the-shelf (commercial) materials could offer a lower risk replacement for AM in delivering LEC to treat corneal diseases.

2 Materials

Prepare all solutions in a class II biological safety cabinet and perform cell culture in a 37 °C, humidified incubator containing a CO_2 supply and a High-Efficiency Particulate Air filter unless specified otherwise.

2.1 Electrospinning PLGA Scaffolds

1. PLGA (50:50) MW: 44 kg/mol (Purac, The Netherlands).
2. Dichloromethane (DCM).
3. 5 mL Syringes.
4. 0.8 mm Blunt end needles.
5. Syringe pump (Genie Plus, Kent Scientific, Torrington, CT, USA).
6. Voltage generator (Genvolt, Bridgnorth, Shropshire, UK).
7. Rotating collector (Heidolph Instruments GmbH, Schwabach, Germany).

2.2 Cell Culture

1. Dulbecco's modified Eagle's medium (DMEM).
2. Ham's F12 medium.
3. Fetal calf serum (FCS).
4. Penicillin and streptomycin.

5. Amphotericin B.
6. Epithelial growth factor.
7. Insulin.
8. Ethylenediaminetetraacetic acid (EDTA) solution.
9. Trypsin–EDTA.
10. Anhydrous EDTA.
11. Phosphate buffer saline (PBS).
12. Medical grade stainless steel donut rings (od 35 mm, id 25 mm).

2.3 Immunocytochemistry

1. Propidium iodide.
2. RNase A from bovine from bovine pancreas.
3. Phalloidin fluorescein isothiocyanate (FITC).
4. 3.7 % (v/v) buffered formaldehyde.

3 Methods

3.1 Electrospinning PLGA Scaffolds

Perform the electrospinning process in a Class II laminar flow cupboard (*see* **Note 1**).

1. Weigh out 5 g of PLGA polymer into an autoclaved bottle. Add 20 g DCM (*see* **Note 2**) to this using a clean glass pipette, to obtain a 20 % (w/w) solution. Stir the solution overnight to ensure complete dissolution.
2. Draw the solution into four 5 mL sterile syringes and fit blunt ended needles (previously soaked in DCM for 10 min to sterilize) onto syringes.
3. Stack the syringes onto the syringe pump and pass each needle through an aluminum plate connected to a voltage supplier (Fig. 1).
4. Set up an earthed rotating collector approximately 20 cm from the tip of the needles, to collect the electrospun sheet as it is extruded from the pumped syringe. The collector should be wrapped in autoclaved aluminum foil then covered with autoclaved polytetrafluoroethylene sheet (Bake-o-glide, Falcon Products Ltd, UK) that can be secured with tape.
5. Set the collector to spin at 300 rpm, run the syringe pump at 50 μL/min and set the voltage supplier to 10–15 kV. As the needles are focussed towards the collector, a jet of polymer emerging from the needle tip will land on the collector by forces of attraction. As the solution spins, a sheet of scaffold will be obtained on the collector. This should be carefully removed and allowed to air-dry overnight. The scaffold can be stored in a sterile petri dish and kept dry until use (*see* **Note 3**).

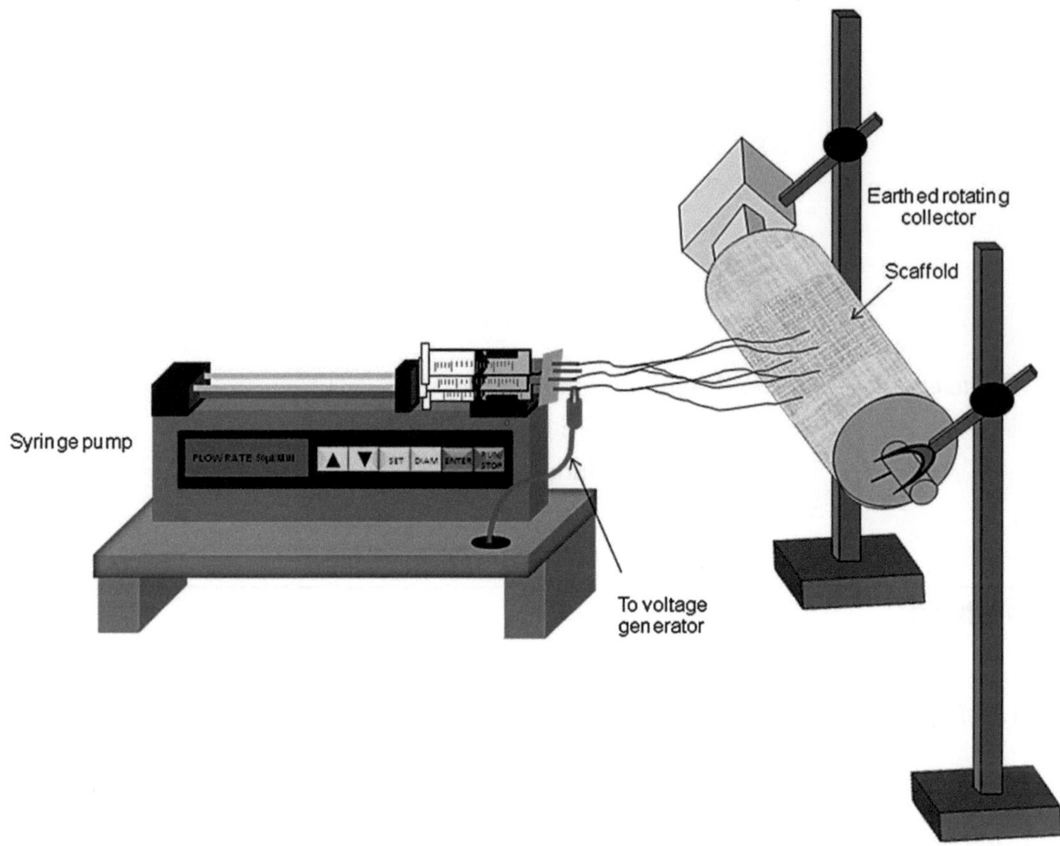

Fig. 1 Setup of the electrospinning rig

3.2 LEC Culture

1. Use DMEM and Ham's F12 medium in a 1:1 ratio supplemented with 10 % (v/v) FCS, 10 ng/mL EGF, 5 μg/mL insulin, 2.5 μg/mL amphotericin B, and 100 IU/mL penicillin and 100 μg/mL streptomycin to culture LEC.
2. Isolate LEC (*see* Chapters 7–9) by excising the limbal rim of corneas into four segments and place those into 2.5 mg/mL of Dispase II (in DMEM) for 45 min. Scrape limbal cells from this tissue into PBS using a pair of blunt forceps. Centrifuge the cell solution at 200 × *g* for 5 min, discard the supernatant and resuspend the cells in fresh media.
3. Seed LEC into T-25 flasks containing growth-arrested 3T3 murine fibroblasts as a feeder layer. Maintain the culture at 37 °C until 80 % confluent.

3.3 Culture of LEC on PLGA Scaffolds

1. Cut PLGA sheets into disks to dimensions which allow them to fit into a 12-well plate (27 mm diameter wells).
2. Place medical grade metal rings of 25 mm diameter on PLGA disks (*see* **Note 4**).

3. Wash LEC three times with sterile PBS when cells are 80 % confluent. Add 1 mL 0.02 % (w/v) EDTA to the flask containing cultured cells and incubate at 37 °C for 3 min to remove 3T3 cells. Then dilute EDTA with 3 mL of PBS, remove solution from the flask and add 1 mL of trypsin–EDTA. Incubate LEC for a further 5 min and inhibit trypsin by adding 3 mL of fresh medium containing 10 % (v/v) FCS. Collect the media containing LEC and centrifuge at 200 × *g* for 5 min to obtain a pellet. Resuspend the cell pellet in fresh medium and count cells using a hemocytometer (*see* Chapter 6).
4. Seed 1×10^5 LEC onto each scaffold (*see* **Note 5**) into the metal ring and culture cells for 2 weeks, changing the media twice a week.
5. At the end of the 2-week culture period, wash LEC three times with PBS and fix cells (at room temperature for 10 min) using 2 mL 3.7 % (w/v) buffered formaldehyde. Aspirate the fixative and wash cells three times with PBS.

3.4 Imaging LEC on PLGA Scaffolds (See Note 6)

1. Add 10 μg/mL of RNase A to fixed LEC on the PLGA scaffold and incubate for 20 min. Aspirate the RNase A from cells, wash three times with PBS and incubate cells in 500 ng/mL Phalloidin FITC at room temperature for 30 min. Then remove this solution and wash the cell/PLGA scaffold three times with PBS.
2. Add 1 μg/mL propidium iodide to LEC on the PLGA scaffold, incubate for 5 min and wash cells three times with PBS.
3. Images of stained LEC can be obtained using a Zeiss LSM 510Meta upright confocal microscope (Fig. 2). Propidium iodide is excited at a wavelength of 536 nm and Phalloidin FITC is excited at 494 nm with an Argon laser. Image analysis can be performed using the LSM browser software.

4 Notes

1. Changes in temperature and humidity may vary the production of electrospun scaffolds. Keeping these parameters constant will result in reproducible scaffolds.
2. Due to the viscosity of DCM, the weight of the solution is considered rather than the volume.
3. PLGA scaffolds are deliberately designed to begin to degrade as soon as they are exposed to any aqueous solution.
4. The weight of the metal ring used for culturing cells keeps PLGA scaffolds from contracting during the period of culture.
5. It is advisable to keep either a glass coverslip or some sterile paper under the scaffold during the course of culture. As the

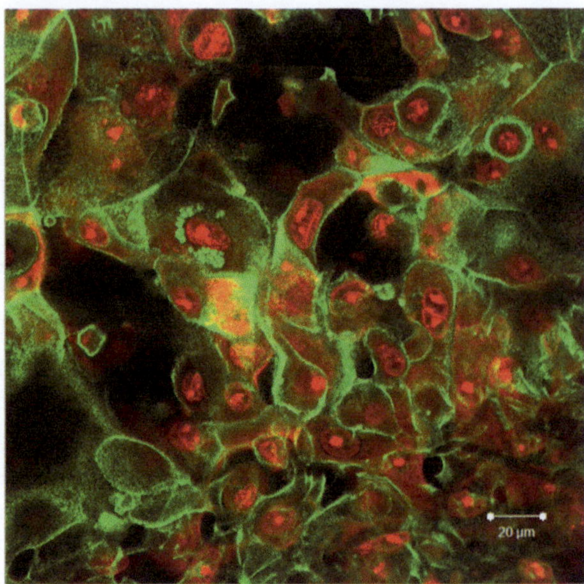

Fig. 2 Confocal microscopy of rabbit limbal epithelial cells cultured for 2 weeks on a PLGA electrospun scaffold. Cell nuclei are stained with propidium iodide (*red*) and actin filaments are stained with Phalloidin FITC (*green*)

scaffolds degrade, they lose their structure and also stick to the bottom of the tissue culture plastic making them difficult to handle.

6. Due to the 3D fibrous structure of PLGA scaffolds, imaging cells on these scaffolds using a standard phase-contrast microscope is challenging. Accordingly, the best way to image the cells is using fluorescence microscopy after staining the cells, for example, by using a live fluorescent cell tracker dye or by immunostaining using antibodies with fluorescent conjugates.

Acknowledgment

This work has been funded by the Wellcome Trust Affordable Healthcare in India Award.

References

1. Gomes JAP, dos Santos MS, Cunha MC, Mascaro VL, Barros Jde N, de Sousa LB (2003) Amniotic membrane transplantation for partial and total limbal stem cell deficiency secondary to chemical burn. Ophthalmology 110:466–473
2. Sridhar MS, Bansal AK, Sangwan VS, Rao GN (2000) Amniotic membrane transplantation in acute chemical and thermal injury. Am J Ophthalmol 130:134–137
3. Sangwan VS, Basu S, MacNeil S, Balasubramanian D (2012) Simple limbal epithelial transplantation (SLET): a novel surgical technique for the treatment of unilateral limbal stem cell deficiency. Br J Ophthalmol 96:931–934
4. Kubo M, Sonoda Y, Muramatsu R, Usui M (2001) Immunogenicity of human amniotic membrane in experimental xenotransplantation. Invest Ophthalmol Vis Sci 42:1539–1546

5. Adinolfi M, Akle CA, McColl I, Fensom AH, Tansley L, Connolly P et al (1982) Expression of HLA antigens, beta 2-microglobulin and enzymes by human amniotic epithelial cells. Nature 295:325–327
6. Endo K, Nakamura T, Kawasaki S, Kinoshita S (2004) Human amniotic membrane, like corneal epithelial basement membrane, manifests the alpha5 chain of type IV collagen. Invest Ophthalmol Vis Sci 45:1771–1774
7. Meller D, Pires RTF, Tseng SCG (2002) Ex vivo preservation and expansion of human limbal epithelial stem cells on amniotic membrane cultures. Br J Ophthalmol 86:463–471
8. Grueterich M, Espana EM, Tseng SCG (2003) Ex vivo expansion of limbal epithelial stem cells: amniotic membrane serving as a stem cell niche. Surv Ophthalmol 48:631–646
9. Sangwan VS, Basu S (2011) Antimicrobial properties of amniotic membrane. Br J Ophthalmol 95:1–2
10. Shimmura S, Shimazaki J, Ohashi Y, Tsubota K (2001) Antiinflammatory effects of amniotic membrane transplantation in ocular surface disorders. Cornea 20:408–413
11. Hao Y, Ma DHK, Hwang DG, Kim WS, Zhang F (2000) Identification of antiangiogenic and antiinflammatory proteins in human amniotic membrane. Cornea 19:348–352
12. Sangwan VS, Basu S, Vemuganti GK, Sejpal K, Subramaniam SV, Bandyopadhyay S et al (2011) Clinical outcomes of xeno-free autologous cultivated limbal epithelial transplantation: a 10-year study. Br J Ophthalmol 95:1525–1529
13. Hopkinson A, McIntosh RS, Tighe PJ, James DK, Dua HS (2006) Amniotic membrane for ocular surface reconstruction: donor variations and the effect of handling on TGF-beta content. Invest Ophthalmol Vis Sci 47:4316–4322
14. Fatima A, Sangwan V, Iftekhar G, Reddy P, Matalia H, Balasubramanian D et al (2006) Technique of cultivating limbal derived corneal epithelium on human amniotic membrane for clinical transplantation. J Postgrad Med 52:257–261
15. Deshpande P, Notara M, Bullett N, Daniels JT, Haddow DB, MacNeil S (2009) Development of a surface-modified contact lens for the transfer of cultured limbal epithelial cells to the cornea for ocular surface diseases. Tissue Eng Part A 15:2889–2902
16. Deshpande P, McKean R, Blackwood KA, Senior RA, Ogunbanjo A, Ryan AJ et al (2010) Using poly(lactide-co-glycolide) electrospun scaffolds to deliver cultured epithelial cells to the cornea. Regen Med 5:395–401
17. Fiorica C, Senior RA, Pitarresi G, Palumbo FS, Giammona G, Deshpande P et al (2011) Biocompatible hydrogels based on hyaluronic acid cross-linked with a polyaspartamide derivative as delivery systems for epithelial limbal cells. Int J Pharm 414:104–111
18. Sudha B, Madhavan H, Sitalakshmi G, Malathi J, Krishnakumar S, Mori Y et al (2006) Cultivation of human corneal limbal stem cells in mebiol gel® - a thermo-reversible gelation polymer. Indian J Med Res 124:655–664
19. Pino CJ, Haselton FR, Chang MS (2005) Seeding of corneal wounds by epithelial cell transfer from micropatterned PDMS contact lenses. Cell Transplant 14:565–571

Chapter 13

The Growth and Delivery of Mesenchymal and Limbal Stem Cells Using Copolymer Polyamide 6/12 Nanofiber Scaffolds

Vladimir Holan, Eliska Javorkova, and Peter Trosan

Abstract

The injured or otherwise damaged cornea is healed by limbal stem cells (LSC). If the limbus where LSC reside is also damaged or nonfunctional, the cornea cannot heal properly and this defect leads to impaired vision that can result in blindness. The only way to treat total LSC deficiency is by transplantation of limbal tissue or a transfer of LSC. Recently, mesenchymal stem cells (MSC) have been shown as another promising source of stem cells for corneal healing and regeneration. Here, we describe a protocol for the use of polyamide 6/12 nanofiber scaffolds for the growth of MSC and LSC, and for their transfer onto a mechanically damaged ocular surface in the experimental mouse model.

Key words Mesenchymal stem cells, Limbal stem cells, Polyamide 6/12 nanofiber scaffolds, Ocular surface reconstruction

1 Introduction

Stem cell (SC) therapy represents the only effective approach to treating total limbal stem cell (LSC) deficiency. Although LSC represent the optimal source of stem cells for ocular surface reconstruction, their use is often disadvantaged by the limited supply of autologous LSC or by a strong immune response if allogeneic cells from unrelated donors are used. Mesenchymal stem cells (MSC) are an alternative and prospective source of SC for corneal surface reconstruction. These cells obtained from bone marrow or adipose tissue can be used as another source of autologous cells. MSC present with good growth properties in vitro, can be differentiated into other cell types and possess immunosuppressive properties which can contribute to the healing process by inhibiting local inflammatory reactions. The first attempts to use MSC for corneal surface reconstruction have been reported [1–3].

One of the problems associated with SC therapy is the absence of a suitable carrier for SC growth and transfer onto the ocular surface.

Bernice Wright and Che J. Connon (eds.), *Corneal Regenerative Medicine: Methods and Protocols*, Methods in Molecular Biology, vol. 1014, DOI 10.1007/978-1-62703-432-6_13,

To date, various supporting materials for the culture, transportation and transplantation of SC onto the recipient eye have been tested. These materials include fibrin glue [4], polymers or collagen sponges [5], and human amniotic membrane (AM) [6]. Promising results have been obtained with these materials, but there is still a need for more versatile carriers of SC for corneal regeneration.

In recent years, promising scaffolds for the growth and transfer of various types of SC have been offered by nanotechnology. Electrospinning processes can fabricate nanofibers with a diameter ranging from a few tens to hundreds of nanometers and with a defined porosity. The three-dimensional structure of nanofibrous materials has an extremely large surface area, and nanofibers can mimic the structure of extracellular matrix proteins, which provide support for cell growth and function. It has been shown that SC grow and differentiate on nanofibers comparably or even better than on plastic surfaces [7–10]. These nanofiber scaffolds prepared from biocompatible materials can serve as carriers of SC for treating ocular surface injuries and LSC deficiency [3, 11].

Here, we describe a protocol for preparation of mouse MSC and LSC and for characterization of their growth on nanofiber scaffolds prepared from copolymer polyamide 6/12 (PA6/12). In addition, we describe the use of nanofiber scaffolds for the transfer of MSC and LSC to treat ocular surface injuries in an experimental mouse model.

2 Materials

Prepare all solutions in a class II biological safety cabinet and perform cell culture in a 37 °C, humidified incubator containing a CO_2 supply and a High-Efficiency Particulate Air filter unless specified otherwise.

1. Dulbecco's modified Eagle's medium (DMEM).
2. Roswell Park Memorial Institute (RPMI) 1640 medium.
3. Media supplements: 10 % (w/v) fetal calf serum (FCS), antibiotics: 100 U/mL of penicillin, 100 mg/mL of streptomycin, and 10 mM 4-(2-hydroxyethyl)-1-piperazineethanesulfonic acid (HEPES) buffer.
4. 0.5 % (v/v) Solution of trypsin from porcine pancreas.
5. Phosphate buffered saline (PBS) (pH 7.4).
6. 0.01 M Ethylenediaminetetraacetic acid (EDTA) (pH 8.0).
7. Nanofiber scaffolds prepared from copolymer polyamide 6/12 (PA6/12).
8. Needleless Nanospider™ machine.
9. Cell sorting buffer for magnetic activated cell sorting (MACS): PBS (pH 7.2) containing 0.5 % (w/v) bovine serum albumin (BSA) and 2 mM EDTA.

10. MACS apparatus (Miltenyi Biotec).
11. CD11b and CD45 MicroBeads (Miltenyi Biotec).
12. Antibodies for flow cytometry: Anti-CD11b (clone M1/70 from BioLegend, San Diego, CA, USA), anti-CD45 (clone 30-F11 from BioLegend), anti-CD44 (clone IM7 from BD PharMingen, San Jose, CA, USA) and anti-CD105 (clone MJ7/18 from eBioscience, San Diego, CA, USA). Use antibodies conjugated with selected fluorochromes according to your flow cytometry facility. Clones of antibodies mentioned in brackets should be optimal for labeling of MSC.
13. Flow cytometer.
14. BALB/c mice.
15. Appropriate animal housing including preoperative and postoperative care. Anesthesia should include xylazine and ketamine.
16. Dissection kit: scissors, scalpels, forceps.
17. Suture material: 11.0 Ethilon (Ethicon, Johnson & Johnson, Livingston, England) and Resolon 7.0 (Resorba, Nuremberg, Germany).
18. Ophthalmic ointment compound containing bacitracin and neomycin (Zentiva, Prague, Czech Republic).
19. Tissue homogenizer.
20. Nylon mesh.
21. Hemocytometer.
22. Percoll solution.
23. Refrigerated centrifuge.
24. CellCrown™24 inserts (Scaffdex, Tampere, Finland).
25. Rubber cell scrappers.
26. 24-Well tissue culture plates.
27. 25-mL Tissue culture flasks.
28. WST-1 cell proliferation assay kit.
29. Spectrophotometer.
30. PKH26 Red Fluorescent Cell Linker Kit.
31. Hoechst 33258 dye.

3 Methods

In our experiments we used nanofibers prepared by a modified needleless Nanospider™ technology [12, 13], in which polymeric jets are spontaneously formed from liquid surfaces on a rotating, spinning electrode. A thin layer of polymer solution film is raised by a metal roller, which simultaneously functions as the positive

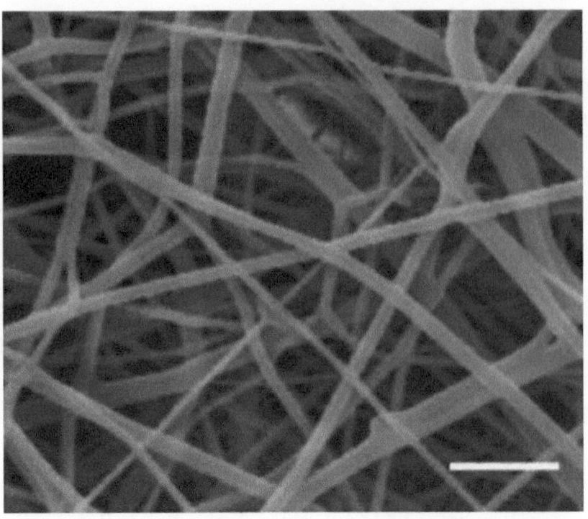

Fig. 1 Scanning electron microscopy of nanofibers fabricated from copolymer PA6/12. Scale bar: 5 μm

electrode. This electrode is partially submerged in the polymer solution, and nanofibers are created between the spinning electrode and a collector due to a very high intensity electrostatic field by the force of Taylor's cones. The parameters of the Nanospider machine during the formation of the PA6/12 nanofiber scaffolds were: 6 rpm for the spinning electrode, 15 cm between the spinning electrode and the collector, 5.5 kV/cm high voltage supply, temperature up to 25 °C, and relative humidity up to 30 %. The solvent evaporates and the fibers stretch at room or elevated temperature.

This Nanospider technology flexibly enables the formation of fibers which are tens of nanometers to tens of micrometers in diameter. All nanofibrous samples were prepared at a basic weight of 3–5 g/m^2 with nanofiber diameter ranging from 290 to 539 nm. The architecture of nanofibers is shown in Fig. 1. The nanofibers can be loaded with various immunosuppressive drugs to inhibit a local rejection or inflammatory reaction after SC transfer [11]. Nanofibers prepared by other special techniques can also be used as scaffolds for SC growth and transfer (*see* **Note 1**).

3.1 Preparation of Nanofibers for Cell Culture

1. Cut nanofiber scaffolds into squares (approximately 1.5 × 1.5 cm) and fix them into CellCrown™24 inserts.
2. Sterilize the inserts containing nanofibers using UV light.
3. Soak inserts with nanofibers overnight in sterile distilled water to avoid possible toxicity caused by residual monomers and organic solvent.
4. Before cell seeding, wash fixed nanofibers with culture medium and transfer them into 24-well tissue culture plates.
5. Add 650 μL of culture medium into each well to soak the insert before adding cells.

3.2 Mesenchymal Stem Cells

3.2.1 Isolation of Bone Marrow Cells

1. For isolation of bone marrow cells we used 8–12 week old BALB/c mice.
2. Sacrifice mice by cervical dislocation and sterilize their surface with 70 % (v/v) ethanol.
3. Remove skin and muscle tissue from hind-limbs; use scissors to expose the femur and tibia.
4. Cut the bones at both ends and transfer them to a beaker containing 5 mL DMEM medium supplemented with 5 % (w/v) FCS.
5. Flush out bone marrow from each bone by rinsing using a 23-gauge needle attached to a 5-mL syringe filled with medium. Collect bone marrow into a test tube and homogenize this using a tissue homogenizer. Filter the cells through nylon mesh and centrifuge the cell suspension at $250 \times g$ for 8 min.
6. Remove the supernatant, add 10 mL of supplemented DMEM medium, and resuspend the pellet into a single-cell suspension.
7. Count cells using a hemocytometer (*see* **Note 2**).

3.2.2 Culture of Bone Marrow Cells

1. Adjust the concentration of bone marrow cells to 2×10^6 cells/mL and culture them in 20 mL of medium in 75-cm^2 tissue culture flask in a humidified atmosphere with 5 % CO_2 at 37 °C.
2. Following a 48-h incubation period, remove non-adherent cells by washing. Wash cells by removing old medium containing non-adherent cells and by adding 10 mL of fresh medium prewarmed for 37 °C. Repeat this step twice. Add 25 mL of fresh prewarmed medium and culture the remaining adherent cells for additional 2–3 weeks (*see* **Note 3**).
3. Every 3–4 days change two-thirds of the culture medium. Using an optical microscope monitor the shape and confluence of growing cells. Small colonies of spindle-shaped cells scattered on the surface of the culture flask can be recognized. These colonies of bone marrow cells expand and reach confluence on days 6–8 (Fig. 2).
4. Passage cells when colonies reach confluence to maintain an optimal cell growth and concentration. Harvest adherent cells from plastic surface by gentle scraping with a rubber scraper and transfer half of the cells into a new culture flask for expansion. After passaging, the cells should adhere, become spindle-shaped, form colonies, and after 6–8 days reach confluence again.
5. Repeat the passage step and expand half of the cells in new culture flasks.
6. Harvest cells immediately after they have achieved confluence in the second passage.

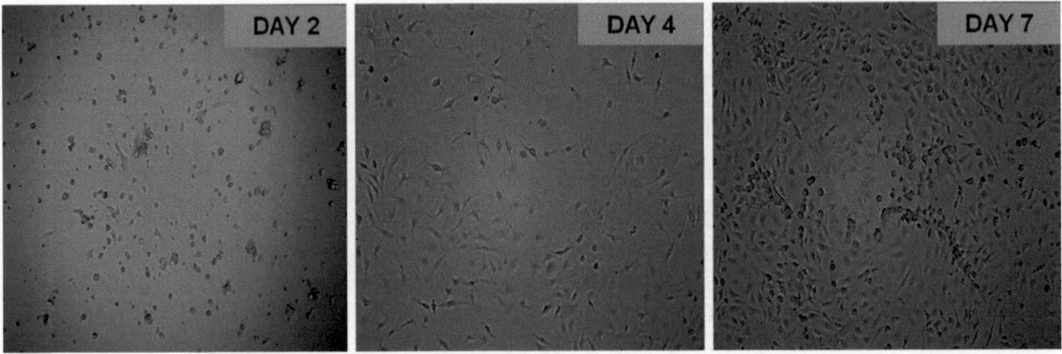

Fig. 2 The growth of bone marrow-derived adherent cells. Mouse bone marrow cells were cultured in plastic tissue-culture flask. Primary round cells start to adhere, proliferate, and reach confluence on day 7. Magnification: 100×

3.2.3 Depletion of CD11b+ and CD45+ Cells

The heterogeneous population of adherent cells growing from bone marrow contains MSC and other cell types which should be removed to isolate MSC. Kits have been designed for the separation of $CD11b^+$ and $CD45^+$ cells from freshly isolated bone marrow cells; some steps in our protocol may differ from the original protocol recommended by the manufacturer. Removal of $CD11b^+$ and $CD45^+$ cells from bone marrow cell cultures is crucial to obtaining a pure population of MSC.

1. Harvest cells by removal of medium, adding 5 mL of 0.01 M EDTA solution for 3 min, shaking and gently scraping.
2. Collect cells into test tubes, centrifuge at $250 \times g$ for 8 min, resuspend the pellet in buffer and count cells using hemocytometer. Centrifuge cells again to remove the remaining culture medium and EDTA solution.
3. Remove the supernatant completely and resuspend the cell pellet in 70 μL of buffer per 10^7 of cells. Add 15 μL of CD11b MicroBeads and 15 μL of CD45 MicroBeads per 10^7 cells. Mix well and incubate for 15 min at 4 °C.
4. Wash cells by adding 2 mL of MACS buffer per 10^7 cells and centrifuge at $250 \times g$ for 8 min. Remove supernatant completely and resuspend pellet in 500 μL of MACS buffer.
5. Perform magnetic separation by using the appropriate MACS Column and MACS Separator.
6. After depletion of $CD11b^+$ and $CD45^+$ cells a cell population that is devoid of $CD11b^+$ and $CD45^+$ cells and that represents approximately 5–10 % of the original cell population isolated from bone marrow should be obtained. This population of purified MSC can be characterized by fluorescent staining of selected surface markers using flow cytometry, and can be subsequently cultured on nanofiber scaffolds (*see* **Note 4**).

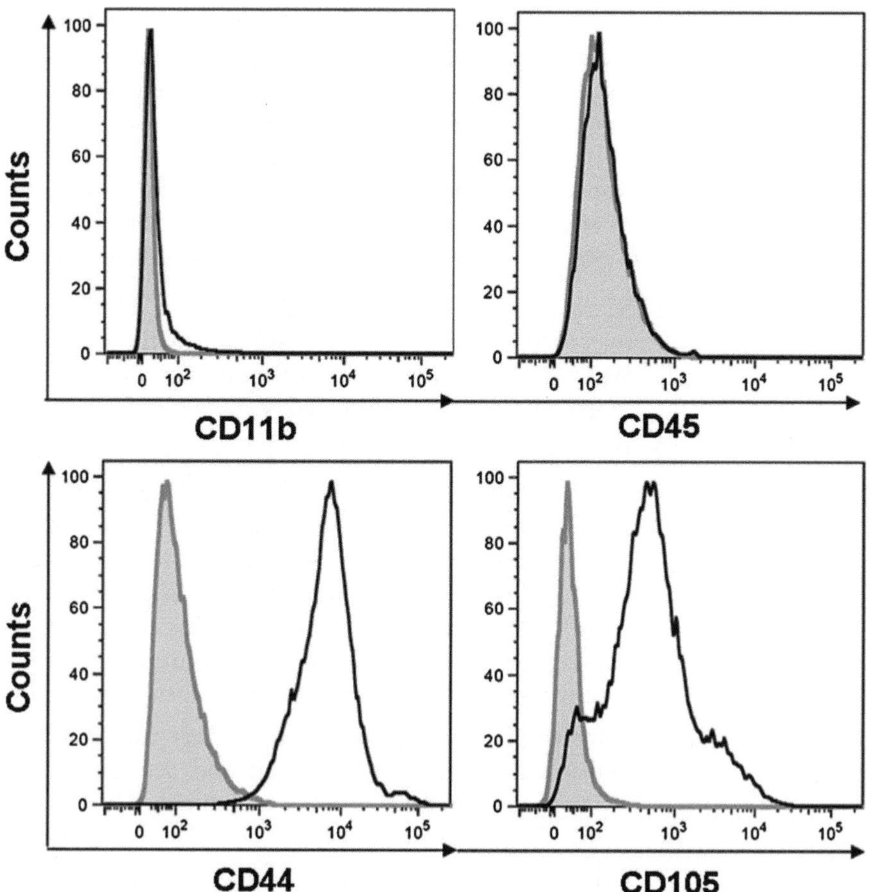

Fig. 3 Phenotypic characterization of sorted bone marrow-derived MSC. Expression of CD11b, CD45, CD44, and CD105 surface markers assessed by flow cytometry. The purified MSC should be negative for CD11b and CD45, and positive for CD44 and CD105

3.2.4 Phenotypic Characterization of MSC by Flow Cytometry

Flow cytometry can be used to validate the purity of sorted MSC through characterization of these cells using positive and negative protein markers. The most common negative markers are CD45 and CD11b, positive markers attributed to murine MSC are CD44 and CD105 (Fig. 3).

1. Wash approximately 0.5×10^6 MSC in PBS containing 0.5 % (w/v) BSA and divide the cell suspension into five samples. Incubate cell samples for 30 min with fluorochrome-labeled monoclonal antibodies anti-CD44, anti-CD105, anti-CD11b, anti-CD45 at 4 °C in a dark chamber. Retain one unlabelled cell sample.
2. Wash each sample with PBS and centrifuge at $250 \times g$ for 3 min. Repeat this step twice.

3. Stain dead cells by adding Hoechst 33258 dye to samples and incubating for 15 min before flow cytometry analysis. Analyze the expression levels of markers. Use the unlabeled cell sample stained only by Hoechst 33258 dye as a negative control.

3.3 Limbal Stem Cells

3.3.1 Isolation of Limbal Cells

LSC can be obtained from human corneoscleral rims as well as a number of animal models including cows, mice, rats, rabbits, mini-pigs, goats and dogs. For this protocol LSC were isolated from mice.

1. Sacrifice mice by cervical dislocation. Under an operative microscope, dissect limbal tissue from the eye using scissors. Wash limbal tissue with PBS and place samples into RPMI 1640 medium containing 10 % (w/v) FCS.
2. Pool limbal tissue from 10 to 12 mice and cut tissue into small pieces in PBS. Centrifuge tissue at 250 × *g* for 8 min.
3. Remove PBS and digest the tissue pellet with 300 μL of 0.5 % (v/v) trypsin for every ten limbal rims. Digest tissue for 10 min at 37 °C. Harvest the supernatant (tissue-free solution) into 20 mL RPMI 1640 medium containing 10 % (w/v) FCS on ice, and repeat the trypsinization procedure on the residual pellet (optimal digestion includes 10–15 trypsinization cycles).
4. After the last trypsinization step, filter the harvested cell suspension through a nylon mesh and centrifuge for 8 min at 250 × *g*. Resuspend the pellet in 1.0 mL RPMI 1640 medium containing 10 % (w/v) FCS.
5. Count cells using a hemocytometer (*see* **Note 5**).

3.3.2 Isolation of LSC Using a Percoll Gradient

The transfer and transplantation of a whole limbal cell population is sufficient to obtain corneal regeneration and healing in the majority of cases. The whole limbal cell population is heterogeneous and contains progenitor and differentiated epithelial cells and stromal keratocytes. LSC represent only a minor population of total limbal cells. Additional purification procedures are necessary to prepare an enriched population of LSC. We have recently described a procedure for isolation of LSC on a discontinuous Percoll gradient [14].

1. Prepare a stock solution by mixing nine parts of Percoll with one part of 10× concentrated PBS. Prepare 40, 50, 60, 70, and 80 % (v/v) Percoll solutions by dilution of the stock solution in PBS (Table 1).
2. Prepare a Percoll gradient in a 10-mL test tube by overlaying 1.0 mL of each Percoll dilution (80 % Percoll at the bottom to 40 % Percoll at the top) (Fig. 4).
3. Overlay the top of the Percoll column with a 1.0 mL suspension of freshly isolated trypsin-dissociated limbal cells (*see* **Note 6**).
4. Centrifuge the gradient with limbal cells for 10 min at 300 × *g* at 4 °C.

Table 1
Preparation of a Percoll gradient

	40 %	50 %	60 %	70 %	80 %
Percoll stock solution (mL)	0.60	0.75	0.90	1.05	1.20
PBS (mL)	0.90	0.75	0.60	0.45	0.30

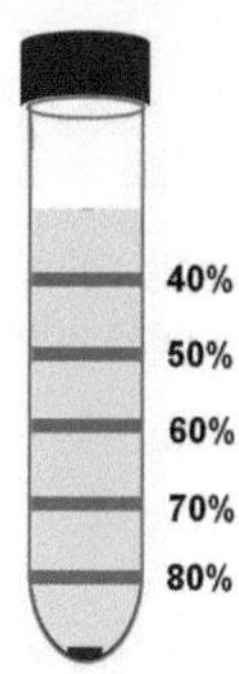

Fig. 4 The scheme of discontinuous Percoll gradient for separation of LSC

5. The separated layers of cells on individual Percoll gradient concentrations can be immediately recognized after centrifugation. Carefully harvest individual cell layers (as well as the cell pellet) into 8 mL RPMI 1640 medium containing 5 % (w/v) FCS and wash three times by centrifugation (8 min at $250 \times g$). LSC are separated in the 80 % Percoll fraction (*see* **Note 7**).
6. After the final wash step, resuspend the cell pellet in 500 μL RPMI 1640 medium containing 10 % (w/v) FCS.

3.3.3 The Growth of Limbal Cells In Vitro

1. Transfer isolated LSC into 25-mL tissue culture flasks at a concentration 1×10^6 cells/5 mL medium/flask. Culture cells in supplemented RPMI 1640 medium with 10 % (w/v) FCS.
2. Replenish half of the medium on the second day of culture with fresh medium.
3. On days 6–8, when the cells reach confluence (Fig. 5), trypsinize the monolayer and transfer cells into 75-cm^2 culture flask to expand growing limbal cells (*see* **Note 8**).

3.4 Growth of MSC and LSC on Nanofibers

1. Transfer MSC or LSC growing in culture flasks into the wells of a 24-well plate containing inserts with fixed nanofibers. Seed 50×10^3 cells per insert/well/700 μL of medium (*see* **Note 9**).
2. Incubate plates for 24 (or 48)h at 37 °C in a CO_2 incubator.
3. Verify the growth and viability of cells growing on nanofibers by the WST-1 assay (*see* **Note 10**). MSC growing on nanofibers prepared from various polymers is shown in Fig. 6.

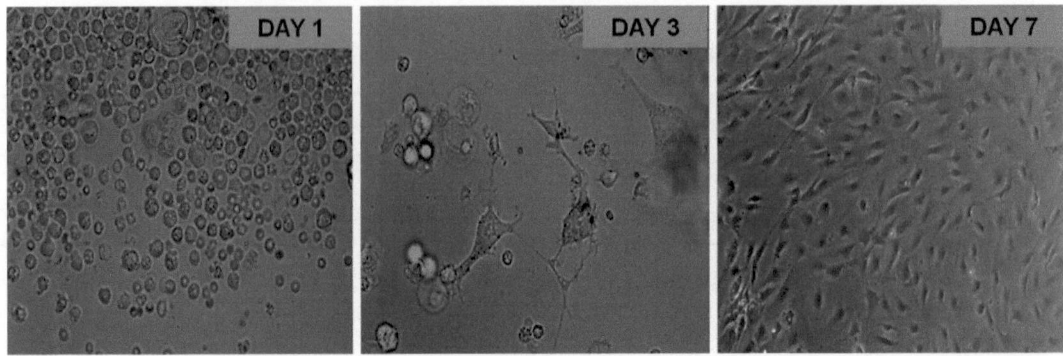

Fig. 5 The growth of limbal cells. Single cell suspension prepared by enzymatic digestion of limbal tissue was seeded in plastic tissue flasks. The cells adhere, proliferate, and reach confluence on day 7. Magnification: 400× (days 1 and 3), 200× (day 7)

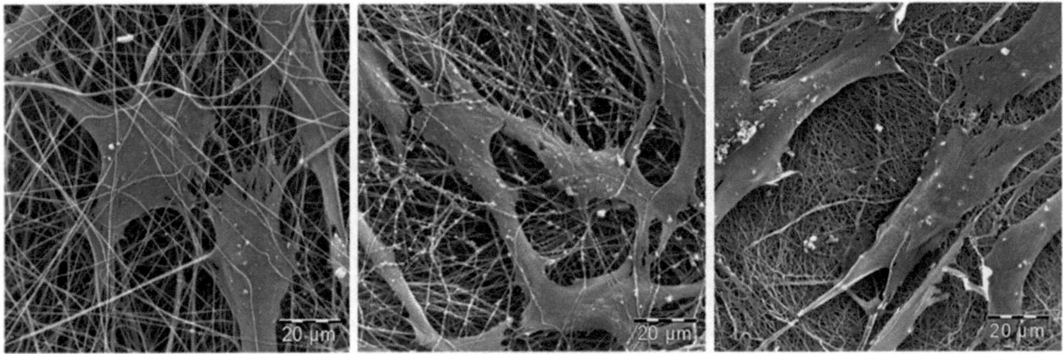

Fig. 6 Morphology of MSC growing on various types of nanofibers. Scanning electron microscopy showed penetration of MSC into the nanofiber structure and the formation of pseudopodia under the surface of nanofibers. Scale bar: 20 μm

3.5 Experimental Model of Corneal Surface Injury

Appropriate ethical approval and facilities must be obtained before undertaking research using animals as experimental models.

1. Deeply anesthetize mice by an intramuscular injection of 80 mg/kg xylazine and 4 mg/kg ketamine.
2. Damage the surface (central cornea region) of the left eye by epithelial debridement with a sharp needle (G23) under an operative microscope. Wash the damaged eye with PBS.

3.6 Transfer of MSC and LSC onto the Damaged Ocular Surface

1. Cut a 4-mm diameter nanofiber circle containing SC (harvested from inserts with growing MSC or LSC).
2. Transfer nanofiber/SC constructs onto the ocular surface with the cell side facing down towards the ocular surface to cover the corneal and limbal region.

3. Suture the nanofiber scaffold containing cells onto the damaged ocular surface by forming four interrupted sutures with 11.0 Ethilon suture material.
4. Close the eyelids for 72 h by tarsorraphy using one suture of Resolon 7.0.
5. Apply an ophthalmic ointment compound containing bacitracin and neomycin onto the ocular surface for 3 days.
6. Remove the nanofiber scaffold from the ocular surface on day 3 after the operation (*see* **Note 11**).
7. Observe the healing process on the ocular surface (*see* **Note 12**).

4 Notes

1. Other examples of processing techniques used to produce nanofibers include drawing out [15], molecular self assembly [16], or thermally induced phase separation [17].
2. The yield of nucleated bone marrow cells from one BALB/c mouse is typically $25–30 \times 10^6$ cells, but this may vary with regard to the strain, sex and age of mice.
3. Always use prewarmed medium. Cold medium could cause detachment of adherent cells.
4. Flow cytometry can be used to confirm the purity of sorted MSC or these cells can be further characterized by assessment of their ability to undergo specific adipogenic and osteogenic differentiation. The protocol for adipogenic and osteogenic differentiation is described in ref. 18.
5. Enzymatic digestion of limbal tissue yields approximately 50,000–100,000 cells per mouse. LSC represent approximately 3–5 % of the limbal cell population.
6. Handle the Percoll gradient column with extreme care to avoid mixing individual Percoll concentrations.
7. The 80 % Percoll fraction contains cells expressing putative LSC markers (ATP-binding cassette sub-family G member 2, p63 and Leucine-rich repeat-containing G-protein coupled receptor 5), and is negative for the corneal epithelial cell marker cytokeratin (CK) 12. The cells also display other characteristics and properties of LSC [14].
8. LSC can be cultured for 3 weeks with a regular passaging or can be frozen in liquid nitrogen for future use. Proliferation of limbal cells can be increased by adding epidermal growth factor (10 ng/mL) or fibroblast growth factor (10 ng/mL) to the culture medium. Insulin-like growth factor 1 (10–100 μg/mL) can support differentiation of LSC into cells expressing corneal epithelial cell-associated marker CK 12 [19].

9. LSC in culture flasks can be released by trypsinization (0.5 % (v/v) solution of trypsin for 5 min) or by a gentle scraping with a rubber scraper.
10. The WST-1 assay is based on the ability of living cells to cleave by mitochondrial dehydrogenases tetrazolium salts into water soluble formazan, which is then measured by spectrophotometry. Culture 50×10^3 cells in 700 μL RPMI 1640 medium in 24-well tissue culture plate with or without inserts containing nanofibers for 24 h at 37 °C in an atmosphere of 5 % CO_2. Add WST-1 reagent (10 μL/100 μL medium) to each well, and incubate the plates for another 4 h to form formazan. Transfer 100 μL formazan-containing medium from each well into the wells of a 96-well tissue culture plate and determine the absorbance using at a wavelength of 450 nm using a spectrophotometer. Always, compare the growth of the same number of cells cultured in inserts with the growth of cells on plastic surface of the 24-well plate.
11. To monitor the migration of cells from nanofibers onto the ocular surface and to trace the fate of transferred cells, MSC and LSC can be stained with the fluorescent vital dye PKH26 or other vital dyes, as we have previously described [3].
12. The therapeutic effect of MSC and LSC delivered to the ocular surface can be observed according to clinical criteria, by the monitoring of transferred cells (if these cells were labeled with some vital dye) or by their effect on local inflammatory reaction, as we have previously described [3, 11].

Acknowledgments

This work was supported by grants P304/11/0653 and P301/11/1568 from the Grant Agency of the Czech Republic, grant KAN200520804 from the Grant Agency of the Academy of Sciences, projects MSM0021620858 and SVV 265211 from the Ministry of Education of the Czech Republic, and project RVO 68378050 from the Academy of Sciences of the Czech Republic.

References

1. Ma Y, Xu Y, Xiao Y, Yang W, Zhang C, Song E et al (2006) Reconstruction of chemically burned rat cornea surface by bone marrow-derived human mesenchymal stem cells. Stem Cells 24:315–321
2. Oh JY, Kim MK, Shin MS, Lee HJ, Ko JH, Wee WR et al (2008) The anti-inflammatory and anti-angeogenic role of mesenchymal stem cells in corneal wound healing following chemical injury. Stem Cells 26:1047–1055
3. Zajicova A, Pokorna K, Lencova A, Krulova M, Svobodova E, Kubínova S et al (2010) Treatment of ocular surface injuries by limbal and mesenchymal stem cells growing on nanofiber scaffolds. Cell Transplant 19:1281–1290
4. Rama P, Bonini S, Lambiase A, Golisano O, Paterna P, De Luca M et al (2001) Autologous fibrin-cultured limbal stem cells permanently restore the corneal surface of patients with total limbal stem deficiency. Transplantation 72:1478–1485

5. Schwab IR, Johnson NT, Harkim DG (2006) Inherent risks associated with manufacture of bioengineered ocular surface tissue. Arch Ophthalmol 124:1734–1740
6. Tsai RJ, Li LM, Chen JK (2000) Reconstruction of damaged cornea by transplantation of autologous limbal epithelial cells. N Eng J Med 343:86–93
7. Xie J, Willerth SM, Li X, Macewan MR, Rader A, Sakiyama-Elbert SE et al (2009) The differentiation of embryonic stem cells seeded on electrospun nanofibers into neural lineages. Biomaterials 30:354–362
8. Xin X, Hussain M, Mao JJ (2007) Continuing differentiation of human mesenchymal stem cells and induced chondrogenic and osteogenic lineages in electrospun PLGA nanofiber scaffold. Biomaterials 28:316–325
9. Shin YR, Chen CN, Tsai SW, Wang YJ, Lee OK (2006) Growth of mesenchymal stem cells on electrospun type I collagen nanofibers. Stem Cells 24:2391–2397
10. Nur-E-Kamal A, Ahmed I, Kamal J, Schindler M, Meiners S (2006) Three-dimensional nanofibrillar surfaces promote self-renewal in mouse embryonic stem cells. Stem Cells 24:426–433
11. Holan V, Chudickova M, Trosan P, Svobodova E, Krulova M, Kubinova S et al (2011) Cyclosporine A-loaded and stem cell-seeded electrospun nanofibers for cell-based therapy and local immunosuppression. J Control Release 156:406–412
12. Jirsak O, Sanetrnik OF, LukasD, Kotek K, Martinova L, Chaloupek J (2005) U. S. patent No. WO 205024101, 2005
13. Dubsky M, Kubinova S, Sirc J, Voska L, Svobodova J, Zajicek R et al (2012) Nanofibers prepared by needleless electrospinning technology as scaffolds for wound healing. J Mater Sci Mater Med 23:931–941
14. Krulova M, Pokorna K, Lencova A, Zajicova A, Fric J, Filipec M et al (2008) A rapid separation of two distinct populations of corneal epithelial cells with limbal stem cell characteristics in the mouse. Invest Ophthalmol Vis Sci 49:3903–3908
15. Xing X, Wang Y, Li B (2008) Nanofibers drawing and nanodevices assembly in poly (trimethylene terephtalate). Opt Express 16:10815–10822
16. Niece KL, Hartgerink JD, Donners JM, Stupp SI (2003) Self-assembly combining two bioactive peptide-amphiphile molecules into nanofibers by electrostatic attraction. J Am Chem Soc 125:7146–7147
17. Huang ZM, Zhang YZ, Kotaki M, Ramakrishna S (2003) A review on polymer nanofibers by electrospinning and their applications in nanocomposites. Compos Sci Technol 63:2223–2253
18. Svobodova E, Krulova M, Zajicova A, Pokorna K, Prochazkova J, Trosan P et al (2012) The role of mouse mesenchymal stem cells in differentiation of naive T-cells into anti-inflammatory regulatory T-cell or proinflammatory helper T-cell 17 population. Stem Cells Dev 21:901–910
19. Trosan P, Svobodova E, Chudickova M, Krulova M, Zajicova A, Holan V (2012) The key role of insulin-like growth factor I in limbal stem cell differentiation and the corneal wound healing process. Stem Cells Dev 21:3341–3350

Chapter 14

Encapsulation and Culture of Mammalian Cells Including Corneal Cells in Alginate Hydrogels

Nicola C. Hunt and Liam M. Grover

Abstract

The potential of cell therapy for the regeneration of diseased and damaged tissues is now widely recognized. As a consequence there is a demand for the development of novel systems that can deliver cells to a particular location, maintaining viability, and then degrade at a predictable rate to release the cells into the surrounding tissues. Hydrogels have attracted much attention in this area, as the hydrogel structure provides an environment that is akin to that of the extracellular matrix. One widely investigated hydrogel is alginate, which has been used for cell encapsulation for more than 30 years. Alginate gels have the potential to be used as 3D cell culture systems and as prosthetic materials, both are applied to regeneration of the cornea. Here, we describe an alginate-based process that has been used for encapsulation of mammalian cells including corneal cells, with high levels of viability, and which allows subsequent retrieval of cell cultures for further characterization.

Key words Alginate hydrogels, Biopolymers, Hydrocolloids, Mammalian cells, Corneal cells, Cell encapsulation

1 Introduction

It is now widely acknowledged that cell delivery could have a beneficial influence on the healing and regeneration of diseased and damaged tissues. While significant research has focussed on the characterization of the cell cultures, including how they respond to different environmental and mechanical stimuli [1], there is relatively little research concentrating on how materials respond to the incorporation of cells into their structure at the molecular level.

Our work has considered both how the cell population may be influenced by their local environments [2] and how they may cause a change in mechanical properties of the gel matrix during culture [3]. Although there are numerous hydrogel-forming biopolymers (chitosan, agarose, carrageenan, gellan) [4], much of the recent work in this area has focussed on the use of calcium alginate for

Bernice Wright and Che J. Connon (eds.), *Corneal Regenerative Medicine: Methods and Protocols*, Methods in Molecular Biology, vol. 1014, DOI 10.1007/978-1-62703-432-6_14, © Springer Science+Business Media New York 2013

encapsulation. The widespread use of alginate is likely because this biopolymer has a relatively long and established history of use as a wound dressing [5], is cheap, and has a very mild gelation mechanism in the presence of calcium ions, which does not compromise the viability of encapsulated cells.

The use of alginate hydrogels for ocular cell therapy is, however, relatively novel. Only a limited number of studies have been performed to examine the manner that alginate may be applied for ocular reconstruction. Alginate microspheres incorporated into collagen hydrogels have been demonstrated as a viable composite construct for human corneal epithelial cell growth [6], and alginate membranes coated with chitosan previously used as base matrices for limbal epithelial cell cultivation maintained the attachment, spreading, and growth of these cells [7]. Recently, calcium alginate hydrogels were shown to be suitable for the transport of limbal epithelial cells under ambient storage conditions [8]. Therefore, prospective uses for alginate hydrogels in corneal regeneration are promising.

Sodium alginate is derived from the brown seaweed (*Phaeophyceae*), and on the molecular scale is a linear polymer formed from two linear uronic acid residues (1-4 linked β-D-mannuronic (M) acid and 1-4 linked α-L guluronic acid (G)). The proportions of these residues and the molecular weight of the polymer can vary depending upon source and the extraction process [9]. Gelation of the alginate hydrocolloid in solution occurs through a process that has been described as "the egg box model" [10]; in this model divalent ions are able to bond between two carboxylate residues found on the G-block of the polymer.

The cross-linking of the polymer chains in the hydrocolloid reduces the mobility of the polymer chains, causing "thickening" of the hydrocolloid and ultimately gelation. While calcium is the cation most often used for alginate gelation in biomedical applications, other divalent metallic ions can also cause gelation to occur. The strength of interactions of divalent ions with the G-block of alginate is as follows: $Pb^{2+} > Cu^{2+} > Ba^{2+} > Sr^{2+} > Ca^{2+} > Mn^{2+} > Mg^{2+}$ [11]. It therefore also follows that it is possible to tailor mechanical properties by using different cross-linking ions. Mechanical properties may also be tailored by using alginates with different molecular weights or M:G block ratios [12]. Whilst forming alginate gel is relatively simple, making samples that are of a desired morphology and are homogeneous without compromising cell viability is more challenging.

This chapter describes how cells may be encapsulated into spherical beads, discs, or other flat-sided shapes of calcium alginate hydrogel for continued cell culture.

2 Materials

Prepare all solutions using Millipore water. We have performed extensive studies using sodium alginate from Sigma, UK, Cat number 180947, lot 08620BJ, MW 102,000–209,000, M:G ratio 1.56, 20–40 centipoise (cps) for 2 % w/v at 25 °C [2, 3, 13]. Other sodium alginates from Sigma can be used, or from other suppliers including NovaMatrix if ultrapure alginates are required. Dispose of all chemicals in a safe manner according to local disposal regulations.

1. High glucose Dulbecco's Modified Eagle's Media (DMEM) with 10 % (w/v) fetal bovine serum (FBS), 1 % (v/v) penicillin–streptomycin solution, 1 % (v/v) fungizone antimycotic liquid (Gibco, UK), 2.25 % (v/v) HEPES solution, and 2 % (v/v) L-glutamine solution.
2. Unsupplemented high-glucose DMEM.
3. Primary or immortalized cells (*see* Chapters 7, 8, and 9 for isolation of corneal cells).
4. Stirrer hotplate and magnetic stirrer.
5. T-flasks between 25 and 75 cm^2 in volume.
6. Phosphate-buffered saline (PBS).
7. 100 mM $CaCl_2$.
8. Hemocytometer or Coulter counters (Beckman Coulter Inc., UK) and trypan blue.
9. P1000 pipette and 1 mL tips.
10. Syringes and syringe needles.
11. 0.2 μm pore size syringe filters.
12. Whatman number 1 filter papers (Fisher Scientific).
13. Sterile Petri dishes.
14. Class II biological safety cabinet.
15. Cell culture CO_2, 37 °C incubator.
16. 100 mM trisodium citrate.
17. Fine forceps.
18. Sodium alginate.
19. Sterile molds of desired dimensions.
20. Trypsin–EDTA solution.
21. Pasteur pipettes.

3 Methods

3.1 Preparation of the Alginate Hydrocolloid

1. Heat the desired volume of water on a stirred hotplate to a temperature of 35 °C (*see* **Note 1**).
2. Weigh out the appropriate mass of sodium alginate salt and add to the water to form the alginate hydrocolloid. This mass should enable preparation of an alginate solution with a concentration between 0.5 and 5 % (w/v) so that a self-supporting, homogeneous hydrogel results (*see* **Note 2**).
3. Very gradually add the sodium alginate salt to the water whilst continuing to heat and stir the hydrocolloid. Be careful not to add the alginate salt too quickly, to prevent agglomerates of the salt from forming within the solution.
4. Continually heat and stir the hydrocolloid until complete dispersion of the alginate salt has occurred.
5. The alginate hydrocolloid can now be sterilized. This can be done by either autoclaving (2 h at 121 °C and 1 bar) or filter sterilizing the solution (through a 0.2 μm pore size syringe filter) (*see* **Note 3**).
6. Store the alginate hydrocolloid at 4 °C until required.

3.2 Preparation of Cell Cultures

A variety of cell types can be encapsulated in calcium alginate hydrogel [14–16]. We have successfully encapsulated and maintained the viability of both NIH3T3 cells (LGC, Middlesex, UK) [2, 3, 12] and primary bone marrow stromal cells [17].

1. Prepare culture media: A simple recipe which can be used for the long-term culture (up to at least 150 days) of a variety of cell types consists of high-glucose DMEM with 10 % (w/v) FBS, 1 % (v/v) penicillin–streptomycin solution, 1 % (v/v) fungizone antimycotic liquid, 2.25 (v/v) HEPES solution, and 2 % (v/v) L-glutamine solution (*see* **Note 4**). Revive and culture cells according to supplier's instructions.
2. Culture cells in supplemented culture media (prepared as described above or according to a preferred recipe (*see* Chapters 7, 8, and 9 for culture of corneal cells)) in tissue culture-treated flasks (T-flasks) of between 25 and 75 cm², depending on the number of cells required. Maintain cultures at 37 °C with 5 % CO_2 and 95 % relative humidity and change media 3–4 times weekly.
3. Passage cultured cells at least twice before their use in experiments, splitting cells at around 70 % confluence at a ratio of between 1/3 and 1/5. This should be done by trypsinization:

   (a) Remove culture media, wash cells in PBS to remove residual media, then add enough trypsin (preheated to 37 °C) to cover the cells, and incubate at 37 °C for a few minutes until cells detach. Detachment can be encouraged by agitating the flask. Add an equal volume of supplemented culture medium to inactivate the trypsin. Pellet the cells by centrifugation at 112 × *g* for 3–5 min.

(b) Remove supernatant and resuspend the cell pellet in fresh culture media before adding cells to a new T-flask.

3.3 Encapsulation of Cells in the Calcium Alginate Hydrogel

All materials should be sterile and procedures should be performed aseptically.

1. Prepare a solution of 100 mM $CaCl_2$ using Millipore water. Sterilize by filtration (0.2 μm pore size filter) or by autoclaving (2 h at 121 °C and 1 bar). Store at 4 °C until required. Heat desired quantities of 100 mM $CaCl_2$ and alginate hydrocolloid to 37 °C. The volume of alginate hydrocolloid required will depend on the number and size of samples you wish to prepare. The $CaCl_2$ volume required will be at least ten times that of the alginate hydrocolloid volume to ensure that there is an excess of $CaCl_2$.
2. To remove cells from T-flasks follow **step 3a** of Subheading 3.2. Before spinning down the cells determine the cell number in the sample, by using either a hemocytometer or a coulter counter.
3. Remove supernatant and add a small volume of alginate hydrocolloid to the cell pellet (*see* **Note 5**). Carefully pipette the solution to dissociate the cell pellet using a P1000 pipette. Very gradually add the remaining amount of alginate hydrocolloid desired to the cell pellet by mixing the solution carefully and continuously using a P1000 pipette to ensure that cell pellet is completely dissociated and the cells are homogeneously distributed throughout the hydrocolloid. A concentration of 7.5×10^5 cells/mL of alginate hydrocolloid allows for the cells to be completely dissociated from each other and for cell viability to be maintained.
4. The alginate/cell dispersion can now be gelled by contact with 100 mM $CaCl_2$. A variety of shapes are possible, including beads and flat-sided shapes. The method to form a disc-shaped gel is shown in Fig. 1.

3.3.1 Calcium Alginate Beads

Add the alginate/cell dispersion dropwise into a bath of 100 mM $CaCl_2$ and incubate at 37 °C with 5 % CO_2 and 95 % relative humidity for 2–3 h to form cross-linked spheres. The diameter of the spheres can be controlled by altering the speed of dropping and the size of the extrusion tip (e.g., a syringe needle or P1000 1 mL tips).

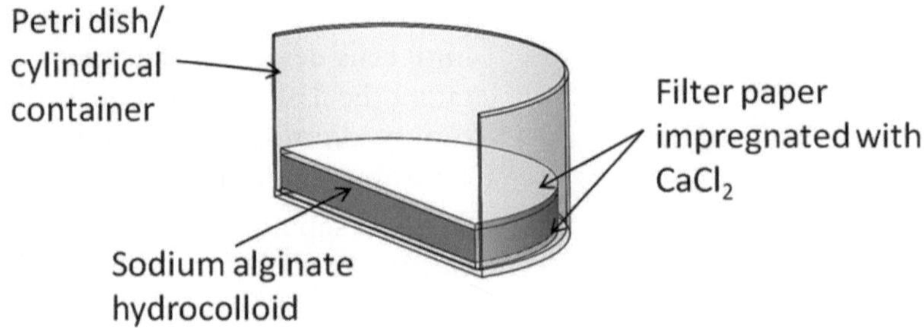

Fig. 1 Schematic illustrating preparation of calcium alginate discs in petri dishes. Place a sheet of 100 mM $CaCl_2$ impregnated number 1 Whatman filter paper on the base of the dish. Cover filter paper with alginate hydrocolloid. Place another sheet of 100 mM $CaCl_2$ impregnated number 1 Whatman filter paper over the surface of the gel, and then add an excess of 100 mM $CaCl_2$

3.3.2 Calcium Alginate Discs and Other Flat-Sided Shapes

1. Place Whatman number 1 filter papers (VWR, UK) in a bath of 100 mM $CaCl_2$ to allow the paper to be impregnated with the $CaCl_2$.
2. Place a piece of the $CaCl_2$-impregnated filter paper on the base of a petri dish. Place sterile open-ended disc or other shaped molds of the desired dimensions (*see* **Note 6**) on the surface of the petri dish (silicone disc molds can be economically and easily prepared by cutting silicone tubing into slices and then sterilized by autoclaving). Alternatively, use a petri dish with the desired diameter of your disc directly.
3. Pipette the alginate/cell suspension into the mold or onto the disc. Carefully place a piece of Whatman number 1 filter paper impregnated with 100 mM $CaCl_2$ on the surface of the alginate/cell suspension to allow the surface of the disc to gel and to prevent displacement of alginate/cell suspension in the following step.
4. Add an excess of 100 mM $CaCl_2$ to the petri dish to immerse the sample (*see* **Note 7**). Incubate the sample at 37 °C with 5 % CO_2 and 95 % relative humidity until complete cross-linking of the alginate has occurred. The time required for cross-linking will depend on the dimensions of the sample. Sample thickness should be limited as cell viability will decrease with increasing time in $CaCl_2$.

3.4 Washing and Culturing of Alginate Hydrogels

All procedures should be done aseptically and vented culture vessels should be used to allow gas exchange. All reagents should be preheated to 37 °C before use.

1. After preparation of alginate hydrogels, remove filter paper and molds (if used for preparation of discs) and wash the samples three times in un-supplemented DMEM to remove residual $CaCl_2$ (*see* **Note 8**).

2. Transfer samples to larger vessels to ensure that sufficient culture media surrounds the samples to maintain cell viability. Add supplemented culture media, as described in Subheading 3.2 above.
3. Incubate samples overnight at 37 °C with 5 % CO_2 and 95 % relative humidity.
4. Remove the culture medium, wash three times in DMEM, and replace with fresh supplemented culture medium.
5. Maintain cultures at 37 °C with 5 % CO_2 and 95 % relative humidity and change medium 3–4 times weekly for maintenance, ensuring that washes are performed with medium rather than PBS.

3.5 Cell Recovery from Alginate Hydrogels

Over time, degradation of alginate hydrogels will occur in normal culture, both in the presence and absence of cells due to the ion exchange of calcium with sodium in the culture media [3]. This degradation results in the release of encapsulated cells. Released cells will adhere to tissue culture-treated surfaces; therefore, cells can be recovered simply by culturing the cell-encapsulating alginate hydrogel in a tissue culture flask or plate. Alternatively, if cell recovery at a faster rate is required this can be achieved by degrading the alginate hydrogel in trisodium citrate:

1. Prepare a solution of 100 mM trisodium citrate in Millipore water. The volume required is approximately five times that of the alginate sample. Sterilize by filtration (0.2 μm pore size filter) or autoclave (2 h at 121 °C and 1 bar). Store at 4 °C until required, when the solution should be preheated to 37 °C.
2. Remove culture medium from the alginate hydrogel samples. Wash in 100 mM trisodium citrate solution and then add an excess of 100 mM trisodium citrate solution.
3. Incubate the gel in 100 mM trisodium citrate at 37 °C until the sample has completely degraded. The time required will depend on the size of the sample and how long it has been maintained in culture (since the gel degrades over time in culture) (*see* **Note 9**). Once the hydrogel has dispersed, centrifuge the resulting solution at 1,000 rpm for 3–5 min to pellet the cells.
4. Remove supernatant and resuspend the cells in supplemented culture medium.
5. Transfer the cells to a T-flask, petri dish, or other tissue culture-treated culture vessel.
6. Maintain the released cell cultures at 37 °C with 5 % CO_2 and 95 % relative humidity and change media 3–4 times weekly for maintenance. The released cells will adhere to tissue culture-treated surfaces. They may, however, have a rounded appearance

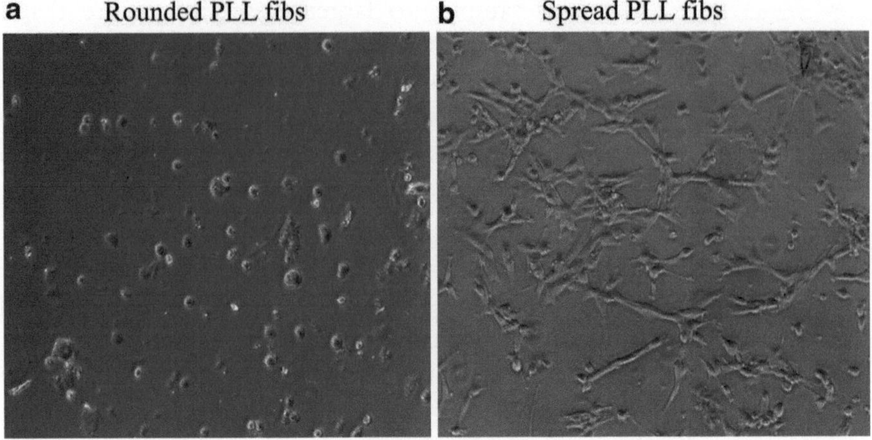

Fig. 2 Appearance of cells after 1-day (**a**) and 7-day (**b**) culture on tissue culture-treated substrates (poly-L-lysine-coated coverslips) after release from calcium alginate hydrogel beads by incubation in 100 mM trisodium citrate for 40 min

during the first few days of culture (Fig. 2a). After a few days, cells should spread and return to normal morphology (Fig. 2b). The time taken for cells to recover normal spread morphology will depend on the time that they were exposed to 100 mM trisodium citrate.

4 Notes

1. Heating the water and stirring continuously will help the alginate to dissolve and help limit the formation of agglomerates.
2. The lower the concentration of the sodium alginate, the softer the resulting gel will be and the quicker the gel will degrade. A higher concentration of sodium alginate will also change the viscosity of the solution, thus causing cells mixed with the hydrocolloid to experience higher shear stress during mixing, which may result in decreased viability of cells following encapsulation. Mechanical testing techniques such as oscillatory rheology can be used to determine the mechanical properties of the hydrocolloid solution and the resulting hydrogel [3].
3. If you choose to autoclave the solution you should take into consideration that degradation of the alginate polysaccharides will occur, resulting in a change in the chemical properties of the alginate and also the mechanical properties of the resulting alginate hydrogel.

4. Store supplemented culture medium at 4 °C and warm only the quantity of medium required to 37 °C before use to prevent degradation of media components.
5. Mix the hydrocolloid thoroughly by pipetting before use in case of sedimentation during storage.
6. Dog-shaped molds can be used to prepare samples for tensile testing in this manner. Samples of desired dimensions can also be cut form a larger disc, such as that made in a petri dish, using cutters such as cork-borers.
7. Petri dishes can be placed inside larger sterile vessels which can be filled with 100 mM calcium chloride solution if required to ensure that an excess of calcium chloride is present to gel the samples.
8. Washing of samples should always be done with medium rather than PBS, as PBS results in greater degradation of the scaffold due to the loss of calcium cross-links from the hydrogel.
9. Incubation time in trisodium citrate should be limited as cell viability decreases with increased exposure time. Incubation times exceeding 3 h should be avoided. If a sample is too large to enable complete dispersion in fewer than 3 h (e.g., a disc) it should be cut into smaller pieces so that degradation is accelerated. Degradation of beads and smaller pieces can be accelerated by agitating the sample with a Pasteur pipette prior to incubation or at internals during the incubation, or by placing the samples on a rocker during the incubation.

Acknowledgments

The work reported in this book chapter was funded by the EU (FP6 NanoBioTact and FP7 NanoBioTouch). The authors would like to thank Michele Marigo (University of Birmingham) for the preparation of Fig. 1.

References

1. Discher DE, Janmey P, Wang YL (2005) Tissue cells fell and respond to the stiffness of their substrate. Science 310:1139–1143
2. Hunt NC, Shelton RM, Grover LM (2009) Reversibel mitotic and metabolic inhibition following the encapsulation of fibroblasts in alginate hydrogels. Biomaterials 30:6435–6443
3. Hunt NC, Smith AM, Gbureck U, Shelton RM, Grover LM (2010) Encapsulation of fibroblasts causes accelerated alginate hydrogel degradation. Acta Biomater 6:3649–3656
4. Van Vlierberghe S, Dubruel P, Schacht E (2011) Biopolymer-based hydrogels as scaffolds for tissue engineering applications: a review. Biomacromolecules 12:1387–1408
5. Gilchrist T, Martin M (1983) Wound treatment with Sorbsan—an alginate fiber dressing. Biomaterials 4:317–320
6. Liu W, Griffith M, Li F (2008) Alginate microsphere-collagen composite hydrogel for ocular drug delivery and implantation. J Mater Sci Mater Med 19:3365–3371
7. Ozturk E, Ergun MA, Ozturk Z, Nurozler AB, Kececi K, Ozdemir N et al (2006) Chitosan-coated alginate membranes for cultivation of limbal epithelial cells to use in the

restoration of damaged corneal surfaces. Int J Artif Organs 29:228–238

8. Wright B, Cave RA, Cook JP, Khutoryanskiy VV, Mi S, Chen B et al (2012) Enhanced viability of corneal epithelial cells for efficient transport/storage using a structurally-modified calcium alginate hydrogel. Regen Med 7:295–307
9. Lee KY, Mooney DJ (2001) Hydrogels for tissue engineering. Chem Rev 101:1869–1880
10. Grant GT, Morris ER, Rees DA, Smith PJC, Thom D (1973) Biological interactions between polysaccharides and divalent cations: the egg-box model. FEBS Lett 32:195–198
11. Haug A, Smidsrod O (1965) Effect of divalent metals on properties of alginate solutions. 2. Comparison of different metal ions. Acta Chem Scand 19:341
12. Wang L, Shelton RM, Cooper PR, Lawson M, Triffit JT, Barralet JE (2003) Evaluation of sodium alginate for bone marrow cell tissue engineering. Biomaterials 24:3475–3481
13. Hunt NC, Shelton RM, Grover LM (2009) An alginate hydrogel matrix for the localised delivery of a fibroblast/keratinocyte co-culture. Biotechnol J 4:730–737
14. Keshaw H, Forbes A, Day RM (2005) Release of angiogenic growth factors from cells encapsulated in alginate beads with bioactive glass. Biomaterials 26:4171–4179
15. Bellamkonda R, Ranieri JP, Bouche N, Aebischer P (1995) Hydrogel-based 3-dimensional matrix for neural cells. J Biomed Mater Res 29:663–671
16. Smith AM, Harris JJ, Shelton RM, Perrie Y (2007) 3D culture of bone-derived cells immobilized in alginate following light-triggered gelation. J Control Release 119:94–101
17. Jahromi SH, Grover LM, Paxton JZ, Smith AM (2011) Degradation of polysaccharide hydrogels seeded with bone marrow stromal cells. J Mech Prop Biomed Mater 7:1157–1166

Chapter 15

In Vivo Confocal Microscopy of the Cornea to Assess Tissue Regenerative Response After Biomaterial Implantation in Humans

Neil Lagali, May Griffith, and Per Fagerholm

Abstract

Laser-scanning in vivo confocal microscopy (IVCM) of the cornea is becoming an increasingly popular tool to examine the living human cornea with cellular-level detail in both healthy and pathologic states. Here, we describe the use of the IVCM technique to examine the processes of tissue healing and regeneration in the living human eye after biomaterial implantation. The regenerative response can be assessed by performing longitudinal IVCM imaging of a laboratory-made, cell-free biomaterial, after direct implantation into a pathologic eye as a primary alternative to human donor tissue transplantation.

Key words Confocal microscopy, In vivo imaging, Cornea, Regenerative medicine

1 Introduction

Laser-scanning in vivo confocal microscopy (IVCM) is the latest generation in a series of clinically approved in vivo confocal microscopes specifically designed to image the living cornea [1–3]. The use of a laser enables a tightly focused spot to be rapidly scanned across an *en face* section of the cornea, to produce high magnification images (up to 600×) at a selectable focal depth in the cornea. The laser enables high-contrast images to be obtained with lateral resolution of 1–2 μm and axial (depth) resolution of 4 μm, in a 400×400 μm field of view, yielding images superior in quality to white-light tandem-scanning or slit-scanning confocal systems [3].

Several comprehensive reviews on the subject of IVCM can be found in the literature [1–3]. The literature describes the use of IVCM in the assessment of the cornea in a healthy state, in pathology, and after various types of surgical or pharmaceutical treatment. Additionally, the technique is being widely adopted outside the clinic, in animal research experiments, and it has been shown

Bernice Wright and Che J. Connon (eds.), *Corneal Regenerative Medicine: Methods and Protocols*, Methods in Molecular Biology, vol. 1014, DOI 10.1007/978-1-62703-432-6_15, © Springer Science+Business Media New York 2013

that corneal morphology is similar across various species [4], which is advantageous for translational research.

As the cornea is normally a transparent tissue, IVCM is capable of providing images of individual cells and nerves in the cornea, which scatter light, and can be readily identified within the transparent extracellular matrix in a dark-field microscope configuration. The IVCM appearance of cells, nerves, and other structures in the cornea has been utilized for assessment and in vivo diagnosis of certain conditions [1–3]. The transparent cornea, however, also provides the opportunity to assess the dynamic nature of corneal cells and nerves, as they participate in regenerative processes.

To explicitly detect and monitor corneal regeneration in vivo, however, requires three components. The first is an actively regenerating tissue microenvironment in the cornea, provided by, for example, a surgically stimulated wound healing response, but without the presence of an interfering inflammatory immunologic rejection reaction that would overwhelm the normal healing process. This requirement is fulfilled by implantation of a biologically compatible biosynthetic matrix constructed of recombinant human collagen (*see* Chapters 9 and 10). The second component is an implanted region that is morphologically distinct from the native recipient cornea, so as to aid in the identification and localization of newly regenerated structures. This requirement is fulfilled by the acellular nature of the biosynthetic material. By IVCM, the implant is easily recognizable as a transparent, cell-free region directly after implantation, whereas surrounding host tissue is readily identified by the presence of cells.

In vivo monitoring of corneal regeneration requires longitudinal observation during multiple patient follow-up visits, over a time period suitable to detect regenerative events. In the cornea, epithelial cell layers regenerate in days to weeks, while stromal keratocytes and nerves can take years to migrate and regenerate. Here, we describe the detailed procedure of performing IVCM in human corneas to detect and qualitatively assess the regenerative response after implantation of an acellular biosynthetic implant.

2 Materials

2.1 In Vivo Confocal Microscope

At present, the only commercially available laser-scanning in vivo confocal microscope for the cornea is the Heidelberg Retina Tomograph 3 with Rostock Cornea Module (HRT3-RCM, Heidelberg Engineering, Heidelberg, Germany). A detailed description of this system, its technical specifications, and its principles of operation can be found elsewhere [5]. The high contrast of this laser-scanning system is critical for detecting fine regenerating epithelial basal cells and nerves, while the high axial resolution enables precise depth localization of individual repopulating keratocytes.

Another commercial system, the white-light slit-scanning Nidek Confoscan4 (Nidek Technologies, Italy), provides in vivo images with a lower lateral and depth resolution.

2.2 Ophthalmic Components for Examinations

1. Ophthalmic gel, a viscous tear substitute in gel form, such as Comfort gel (Bausch & Lomb), Tear-Gel (Novartis), or Viscotirs Gel (CIBA Vision).
2. Sterile single-use PMMA disposable cap, TomoCap®, available from Heidelberg Engineering. Purchased in boxes of 50 caps.
3. Anesthetic eye drops in sterile single-use packaging, such as oxybuprocaine hydrochloride 0.4 % (w/v) or tetracaine hydrochloride 1 % (w/v).
4. A metal wire eyelid speculum (*see* **Note 1**).
5. Long-handle cotton-tipped applicators/swabs (*see* **Note 2**).
6. 70 % (v/v) ethanol solution for cleaning.

3 Methods

Carry out the following procedures in a dimly lit or semi-darkened room.

3.1 HRT3-RCM Instrument Preparation

1. Insert the hardware key into a free USB port, and launch the Heidelberg Eye Explorer software.
2. Ensure that the correct field lens (400 μm) is inserted into the RCM, as per manufacturer's instructions. Use of this field lens will yield a 400×400 μm field of view, which is optimal for patient examinations.
3. Clean and disinfect the chin and forehead rest with ethanol solution.
4. Gently wipe clean the front surface of the objective lens with a lint-free tissue or gently clean with a small amount of distilled water on a cotton swab, or periodically, with a mild liquid detergent.
5. Place a large drop (about 5 mm in diameter) of the ophthalmic gel on the front lens surface of the microscope objective lens (*see* **Note 3**).
6. Remove the TomoCap® from the sterile packaging and securely place it over the gel-covered objective lens. Ensure that the gel is flattened over the front surface of the objective lens by the TomoCap®, and that no air bubbles are present in the gel. Check to ensure the TomoCap® front surface is perfectly horizontal and parallel to the objective lens.
7. Place a drop of ophthalmic gel on the front surface of the TomoCap®, about 5 mm superior to the central region.

The viscous gel will flow downward towards the central region automatically in 1–2 min.

8. The objective lens is mounted inside the RCM housing, which is rotated manually by hand to change the depth of the focal plane inside the cornea (*see* **Note 4**). In this manner, manually adjust focal depth to locate the bright specular reflection peak in the image acquisition window, which corresponds to the outer surface of the TomoCap®. Continue to turn the housing a short distance (typically 10–20 μm, towards larger positive numbers on the depth display) such that the specular reflection is no longer visible and the image acquisition window is dark.
9. Reset the depth to "0" at this point by clicking on the "Reset" button.

3.2 Ex Vivo Imaging of Biomaterial Samples

1. Mount the sample on a glass slide or similar surface using for example double-sided tape.
2. Mount the sample in a vertical orientation, just below the forehead rest of the instrument, to simulate the position of a human eye.
3. Start the examination in the software, such that the image acquisition window is visible. Set the acquisition type to "Sequence" and the acquisition rate (frame rate) to the desired value (*see* **Note 5**).
4. Slowly position the gel-coated TomoCap® in the central region of the biomaterial, using the manual lateral and axial control knobs on the HRT3.
5. Use the side camera image on the image acquisition window to aid in bringing the sample in contact with the gel. The sample should be moved through the gel to a position just outside the TomoCap®.
6. When the sample is visible in the image acquisition window, initiate acquisition by pressing down the foot pedal.
7. During acquisition (a scan), carefully adjust focal depth while images are being acquired (*see* **Note 6**).
8. Repeat **step 7** above for subsequent scans. The lateral and axial position for scanning can be changed at any time during or between scans, to examine different regions of the sample (*see* **Note 7**).
9. Images of the biomaterial sample can be viewed in real time or after scans have been completed (*see* Fig. 1).

3.3 Patient Preparation

1. After instrument preparation, instill one or two drops of topical anesthetic to the eye to be examined (*see* **Note 8**).
2. Ask the patient to place their chin on the chin rest, and forehead against the forehead rest. The patient should then be instructed to look at the white fixation light.

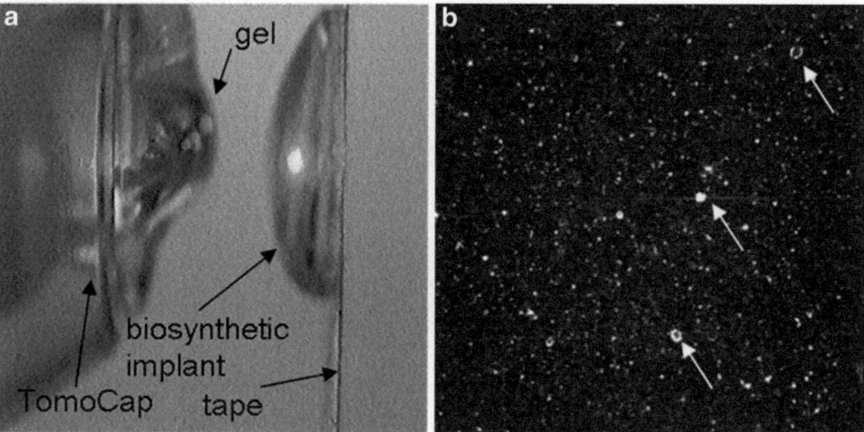

Fig. 1 (**a**) Placement of biomaterial sample for ex vivo microscopic analysis. The biosynthetic implant must make contact with the TomoCap® surface for proper imaging. (**b**) Image taken at a depth of 100 μm inside the biomaterial, with air bubbles (*arrows*) visible, along with smaller, micron-scale inhomogeneities. The image in (**b**) is 400 × 400 μm

3. Adjust the fixation light position such that the desired region of the cornea to be examined is facing the TomoCap®.
4. Coarsely align the eye with the TomoCap® by turning the chin rest height adjustment control and the RCM height adjustment knob.
5. When the eye and TomoCap® are roughly aligned, move the TomoCap® closer to the cornea with the RCM axial alignment knob, but ensure the eyelashes are outside the gel.
6. Instruct the patient to blink, and then open the eye as wide as possible.
7. The TomoCap® and eye should be visible in the live camera window in the software. Use the camera image and the naked eye while adjusting the axial alignment to bring the cornea inside the gel (*see* **Note 9**).
8. Once the cornea is inside the gel, turn the RCM housing to a depth of about 50–60 μm. Slowly move the TomoCap® closer to the cornea by adjusting the axial alignment, until an image is seen in the image acquisition window.

3.4 Image Acquisition

1. The live image seen in the acquisition window may not be in the desired corneal region. Adjust the axial and lateral alignment controls to give an *en face* image of the desired region, for example the central cornea (*see* **Note 10** and Fig. 2).
2. Once the desired corneal region is reached, press the foot pedal to begin image acquisition.
3. While images are being acquired, adjust focal depth to the desired corneal layer. For a full-thickness scan, start acquisition

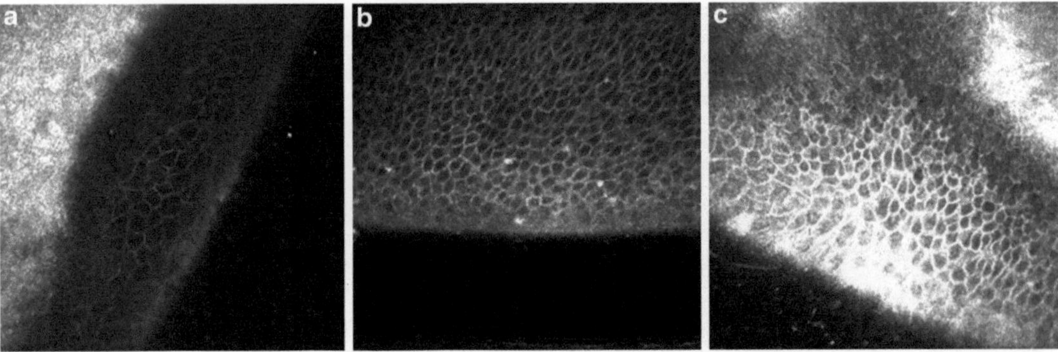

Fig. 2 Oblique in vivo images of the epithelium obtained prior to adjustment of position to achieve *en face* images. The orientation and curvature of the structures in the images indicates their positioning in the cornea. In (**a**), the epithelium in the mid-peripheral cornea is imaged obliquely at a 4 o'clock position. In (**b**), the epithelium is imaged obliquely at a 6 o'clock position. In (**c**) the oblique image indicates roughly the 7 o'clock position

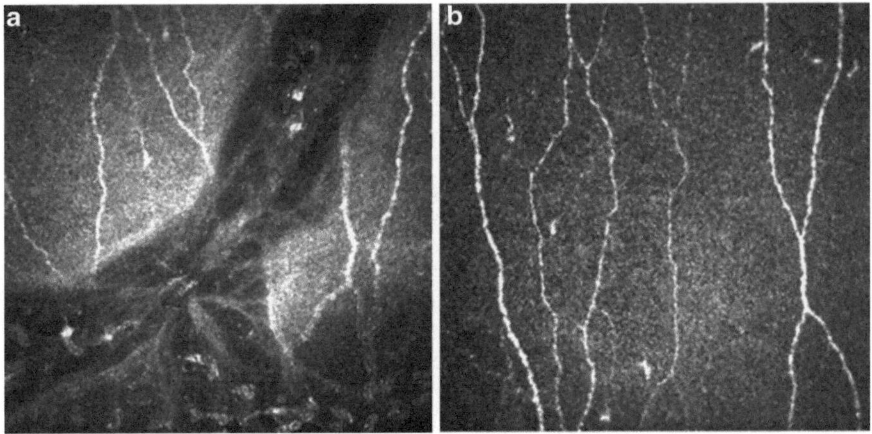

Fig. 3 Image artifacts caused by pressure placed on the cornea by the TomoCap®. (**a**) The sub-basal nerve layer appears disconnected and the underlying anterior stroma is visible with a "folded" or "wrinkled" appearance. (**b**) The same corneal region after releasing pressure on the cornea by adjusting the coarse axial alignment slightly. Sub-basal nerves are visible in a single depth plane, without the bright haze of Bowman's layer or anterior keratocytes visible

at the superficial epithelium and adjust focal depth posteriorly during acquisition (*see* **Note 11**).

4. During image acquisition, pay careful attention to the contact between the cornea and the TomoCap®. The surfaces should just touch one another. If the TomoCap® is too close, pressure will be placed on the cornea and result in image artifacts (folded or wrinkled appearance, *see* Fig. 3).

3.5 In Vivo Imaging: Epithelium

Images of the regenerated epithelial cell layers can be obtained *en face* or obliquely, to study the regeneration process. *En face* images are useful for studying cell morphology and cell density, while

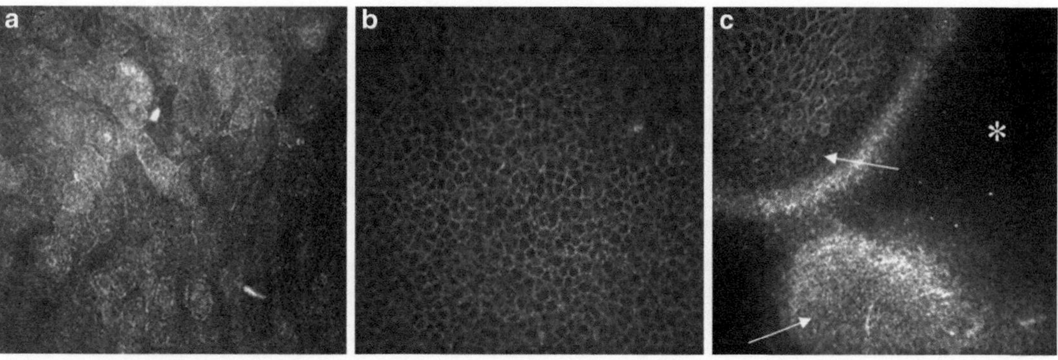

Fig. 4 Regenerated epithelial cell layers covering biosynthetic implants, 3 years after implantation in patients. (**a**) Regenerated, differentiated superficial epithelial cells covering the central implant. (**b**) Regenerated wing cell layer at a mid-epithelial depth. (**c**) Regenerated basal epithelial cells (*arrows*) immediately anterior to a cell-free implant region (*asterisk*). Note that the basal epithelium appears discontinuous in the *en face* image, due to the uneven surface of the anterior implant

oblique images enable a more general analysis of integration of the biomaterial within the host.

1. For *en face* images, adjust focal depth to bring into focus the outermost superficial epithelial cell layer (*see* Fig. 4). Acquire images, while adjusting lateral alignment so as to capture a broad central region of the epithelial cells.
2. Adjust focal depth posteriorly to bring the wing cell layer into focus and repeat the previous step.
3. Finally repeat the previous step for basal epithelial cells. The basal cells directly overlying the biomaterial can be an undulating layer, appearing discontinuous in *en face* sections.
4. For oblique images, adjust the lateral position of the TomoCap® peripherally such that all epithelial cell layers and underlying biomaterial are visible in a single frame (*see* **Note 12** and Fig. 5).
5. Imaging epithelial regeneration in a biomaterial implant requires repeated scans of the epithelium in central, mid-peripheral, and peripheral corneal regions.

3.6 In Vivo Imaging: Sub-basal Nerves

Regenerating sub-basal nerves may not be present in a biomaterial implant for up to several years after implantation. It is recommended to image the peripheral cornea first, to locate the peripheral sub-basal nerves of the recipient. In the peripheral cornea, the interface between implant and host tissue is usually defined by bright, hyper-reflective scar tissue from the healing surgical wound. When sub-basal nerves are found, it is recommended to perform multiple scans to obtain the greatest number of sub-basal nerve images possible, for subsequent selection and analysis.

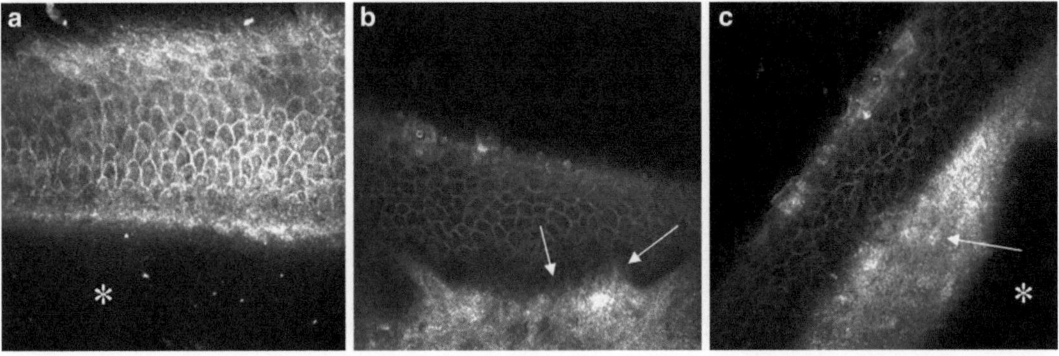

Fig. 5 Oblique in vivo sections of regenerated epithelium overlying biosynthetic implants 3 years after implantation. (**a**) Stratified epithelial cell layers are visible, with a sharp, smooth interface to the underlying biosynthetic implant (*asterisk*). (**b**) Case of regenerated epithelium over an uneven implant anterior surface marked by fibrous tissue (*arrows*). Note that the wing and basal epithelial cell layers are not confined to a single depth plane. (**c**) Regenerated epithelial cell layers cover a stromal region containing repopulated keratocytes (*arrow*), with a cell-free region of the biosynthetic implant visible underneath (*asterisk*)

1. Locate the hyper-reflective scar tissue in a peripheral corneal region, using the fixation target to direct the eye of the patient to an appropriate location.
2. Adjust the focal depth in an anterior direction until the epithelial layers are reached (epithelial wing cells will be visible on one or both sides of the scar).
3. Slowly adjust the focal depth in an anterior or posterior direction to locate the basal epithelium, and the sub-basal nerves that are present at this depth.
4. Acquire images of the peripheral recipient nerves entering the scar tissue region, and adjust lateral and axial alignment to follow these nerves through the scar tissue and into the implanted region (*see* Fig. 6).
5. Once the sub-basal nerves in the implanted region are visible (often with a hazy background of basal epithelial cells), adjust the lateral and axial controls to trace the path of these nerves into the central cornea (*see* **Note 13**).

3.7 In Vivo Imaging: Stroma, Keratocytes, and Endothelium

1. Adjust the lateral and depth positioning to a peripheral stromal region, in order to detect the hyper-reflective scar tissue boundary between implant and host tissue.
2. Once the boundary has been found, trace the path of this boundary around the implant region (*see* **Note 14** and Fig. 7).
3. Simultaneously adjust the focal depth to visualize stromal myofibroblasts or keratocytes in the vicinity of the boundary, giving special attention to cells appearing to migrate into the implant (*see* Fig. 8).

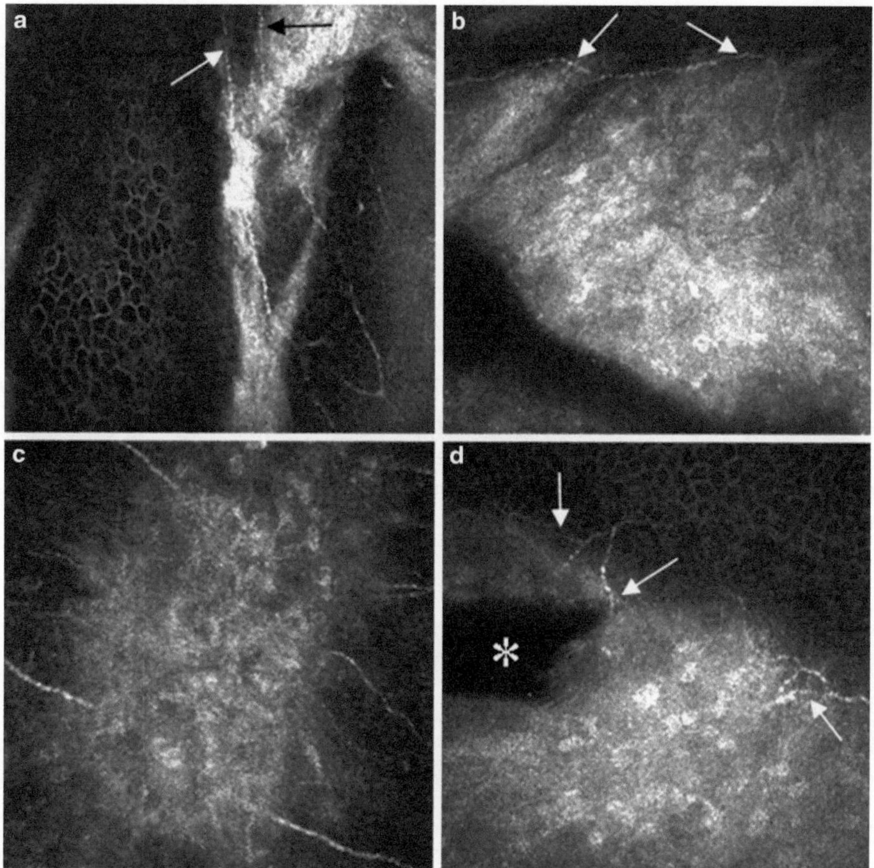

Fig. 6 Regeneration of peripheral sub-basal nerves in a biosynthetic implant 2 and 3 years after implantation. (**a**) Nerves (*arrows*) traversing scar tissue at 2 years. (**b**) In another patient at 3 years, the path of regenerating nerves appears to follow the boundary of reflective scar tissue (*arrows*). (**c**) Nerves traverse a scar region in a patient at 3 years. (**d**) Oblique image of peripheral nerves (*arrows*) entering a region of diminished tissue reflectivity (individual cells visible) in the vicinity of a cell-free region of the biosynthetic implant (*asterisk*) at 3 years

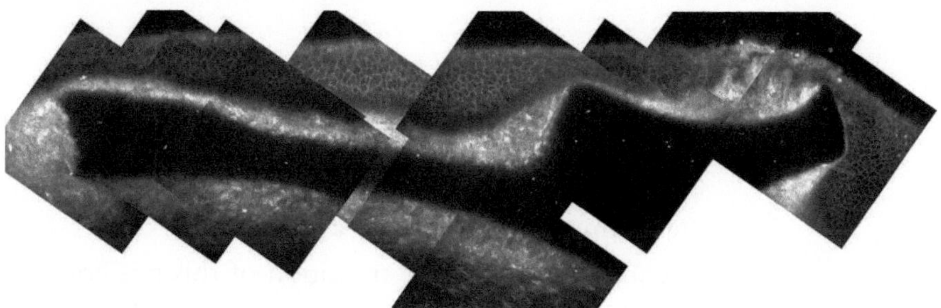

Fig. 7 IVCM montage of the engraftment of biosynthetic material in a patient cornea 18 months after implantation. The central biosynthetic implant appears dark, while peripheral interface regions are cell-invaded. All interfaces contained light-scattering tissue with fibroblasts that serve to anchor the implant in place. A stratified epithelium covers the implant. Note that the image appears slightly distorted due to corneal curvature and the oblique imaging technique used. Note the fine particulate features within the implant in vivo, also detected ex vivo in Fig. 1b

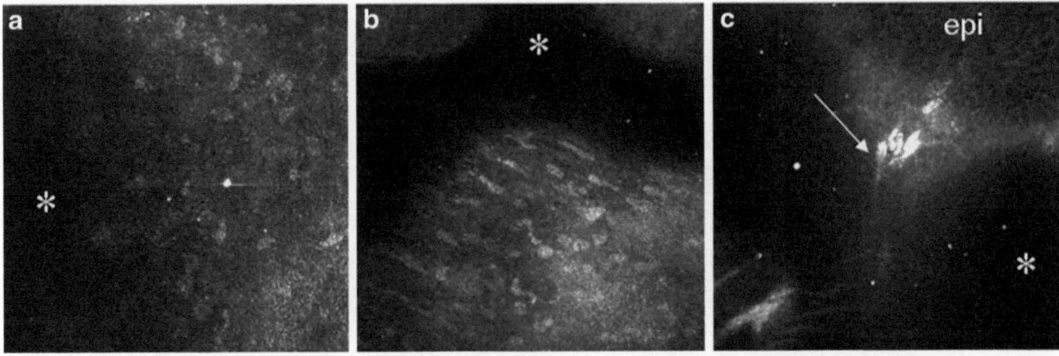

Fig. 8 Stromal keratocyte migration into the biosynthetic implant after 3 years. Dark areas (*asterisks*) represent unpopulated cell-free areas of the implant. Cells migrate into the biomaterial from posterior (*left image*), peripheral *(middle image*), and anterior (*right image*) regions. Note the discrete cells invading the extracellular matrix (*arrow*) in the subepithelial space (epithelial cells marked by "epi"). Note again the fine particulate features in the implant material

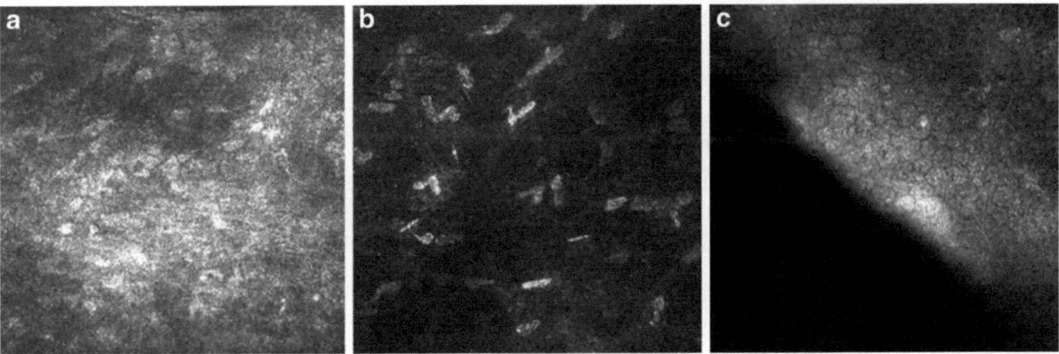

Fig. 9 Imaging the posterior recipient bed underlying the implanted region. (**a**) Keratocytes or myofibroblasts in the posterior interface region 4 years after surgery. (**b**) Native, quiescent keratocytes in the stromal bed of the recipient 4 years after surgery. (**c**) Oblique image of the endothelial cell layer in a patient 2 years after surgery

4. Locate the hyper-reflective posterior implant-to-host interface. This interface will be defined by scar tissue, with stromal keratocytes immediately posterior to the interface, and the biomaterial implant immediately anterior.
5. Acquire images of the transition of this posterior interface to the implanted region, paying particular attention to any cells that appear to migrate into the implant.
6. To check the viability of the posterior recipient stromal bed and endothelium, acquire images while adjusting the focal depth posteriorly from the posterior implant-to-host interface, until the endothelium is reached (*see* Fig. 9).

4 Notes

1. The eyelid speculum is used in cases where the eyelid obscures the cornea, the patient is unable to open the eyelids sufficiently to access the desired corneal region for examination, and the eyelids have a tendency to shut upon exposure to the laser light of the IVCM. In these cases, use a speculum in one eye at a time, and thoroughly anesthetize the eye with at least three drops prior to insertion of the speculum.
2. In some cases, the speculum can be avoided and it is sufficient to manually hold the upper eyelid open with a cotton-tipped applicator. Usually, the upper eyelid is held open in this manner for a period of time sufficient for the gel-covered TomoCap® to make contact with the cornea. The eyelid can thereafter be released, as the TomoCap® serves to hold the upper eyelid open.
3. The RCM is supplied with a 63× 0.95 NA water-immersion objective lens (Zeiss), suitable for use with transparent ophthalmic gel. If another lens is used, ensure that it is of the water immersion type.
4. Changing focal depth manually during an examination results in operator-induced motion artifacts that can limit the quality and quantity of useable images obtained. To overcome this problem, an automatic motor-driven depth controller is now offered as an upgrade to the RCM (Motor Z-Scan Drive System). Note that with the motorized version, a joystick and button replaces the manual rotation and foot pedal, respectively.
5. Acquisition rate can be selected from 1 to 30 FPS, with the number of images fixed at 100 sequential images for each individual sequence scan. Typically, 8 FPS suffices for most patient examinations. In cases where the eye is prone to move during examination, this rate can be increased to 10 or 15 FPS.
6. During image acquisition, a blue bar will appear at the bottom of the window, indicating the elapsed time of the scan. As soon as a scan finishes, a new scan can be started by pressing the foot pedal. The focal depth is recorded along with each image for later review.
7. With patient examinations, the eye can move during a scan, so dynamic adjustment of lateral and/or axial controls is often required to realign the TomoCap® to the desired corneal location. This is a process that requires patience, practice, and experience.
8. It is best to administer a local anesthetic to each eye directly prior to examination. The anesthetic frequently produces a sharp stinging sensation that subsides after 10–15 s. The drops can stimulate tear secretion and a tissue can be given to patients

to dry the area around the eye. It is important that the patient does not rub or disturb the cornea, since the protective blink reflex is suppressed by the anesthetic.

9. If the patient blinks and sweeps away the gel, instruct the patient to remove their head from the rest. Then wipe the front surface of the TomoCap® and place another drop of ophthalmic gel on the TomoCap®, and reattempt alignment.
10. Use the obliqueness of the image as an indicator of the direction in which to move to reach the desired corneal location (*see* Fig. 2).
11. If the acquisition (100 sequential images) finishes before the endothelial cell layer is reached, reduce the frame rate and repeat the acquisition.
12. Oblique sections can be achieved by either adjusting lateral positioning of the TomoCap® for mid-peripheral sections, or adjusting both lateral position and the eye fixation target for peripheral sections.
13. Because the implant anterior surface may not be smooth, the regenerated basal epithelial cell layer, and consequently, the sub-basal nerve plexus, will likely be undulating. This will make imaging of this layer difficult in two dimensions. To overcome this problem, it is recommended to reconstruct sub-basal nerve paths in three dimensions, using motion-free image stacks (obtained from a sequence scan) provided by the Z-Scan Drive System (*see* **Note 4**). Three-dimensional reconstruction from the image stacks can be achieved using commercial software [6]. It is recommended that these image stacks serve as the starting point for sub-basal nerve density quantification, to avoid possible errors in measurement of undulating nerves from two-dimensional images.
14. At early postoperative times (for example, during the first two postoperative years), the transition from scar tissue to implant may be abrupt, with the scar tissue appearing hyper-reflective, and the implant appearing dark and without cells. Over time, however, as keratocytes repopulate the implant region and modify the extracellular matrix, the transition from implant to host may become more gradual and difficult to discern.

Acknowledgments

The authors wish to acknowledge the contribution of Kimberley Merrett (University of Ottawa Eye Institute) and Yuwen Liu (CooperVision), for developing and fabricating, respectively, the biomaterial samples for *ex vivo* imaging; and the Canadian Stem Cell Network and NSERC Canada for grant funding to May

Griffith for biomaterials development. The authors also thank Professor Joachim Stave, University of Rostock, for discussions concerning the Z-Scan Motor Drive for the Rostock Cornea Module.

References

1. Patel DV, McGhee CNJ (2007) Contemporary in vivo confocal microscopy of the living human cornea using white light and laser scanning techniques: a major review. Clin Experiment Ophthalmol 35:71–88
2. Guthoff RF, Zhivov A, Stachs O (2009) In vivo confocal microscopy, an inner vision of the cornea – a major review. Clin Experiment Ophthalmol 37:100–117
3. Niederer RL, McGhee CNJ (2010) Clinical in vivo confocal microscopy of the human cornea in health and disease. Prog Retin Eye Res 29:30–58
4. Reichard M, Hovakimyan M, Wree A, Meyer-Lindenberg A, Nolte I, Junghans C et al (2010) Comparative in vivo confocal microscopical study of the cornea anatomy of different laboratory animals. Curr Eye Res 35:1072–1080
5. Guthoff RF, Baudouin C, Stave J (2006) Atlas of confocal laser scanning in-vivo microscopy in ophthalmology. Springer, Heidelberg
6. Stachs O, Zhivov A, Kraak R, Stave J, Guthoff R (2007) In vivo three-dimensional confocal laser scanning microscopy of the epithelial nerve structure in the human cornea. Graefe's Arch Clin Exp Ophthalmol 245:569–575

Index

Bernice Wright and Che J. Connon (eds.), *Corneal Regenerative Medicine: Methods and Protocols*, Methods in Molecular Biology, vol. 1014, DOI 10.1007/978-1-62703-432-6, © Springer Science+Business Media New York 2013

D

E

F

G

H

I

K

L

T

V

W

X

MIX
Papier aus verantwortungsvollen Quellen
Paper from responsible sources
FSC® C105338

If you have any concerns about our products,
you can contact us on
ProductSafety@springernature.com

In case Publisher is established outside the EU,
the EU authorized representative is:
Springer Nature Customer Service Center GmbH
Europaplatz 3, 69115 Heidelberg, Germany

Printed by Libri Plureos GmbH
in Hamburg, Germany